Bioenergy and Economic Development

Other Titles Published in Cooperation with the Center for Strategic and International Studies, Georgetown University

International Security Yearbook 1984/85, edited by Barry M. Blechman and Edward N. Luttwak

NATO—The Next Generation, Robert E. Hunter

Modern Weapons and Third World Powers, Rodney W. Jones and Steven A. Hildreth

The Emerging Pacific Community, edited by Robert L. Downen and Bruce J. Dickson

The Cuban Revolution: 25 Years Later, edited by Hugh S. Thomas, Georges A. Fauriol, and Juan Carlos Weiss

The U.S. and the World Economy: Policy Alternatives for New Realities, edited by John Yochelson

Banks, Petrodollars, and Sovereign Debtors: Blood from a Stone, Penelope Hartland-Thunberg

Forecasting U.S. Electricity Demand: Trends and Methodologies, Adela Bolet

U.S.-Japanese Energy Relations: Cooperation and Competition, edited by Charles K. Ebinger and Ronald A. Morse

National Security and Strategic Minerals: An Analysis of U.S. Dependence on Foreign Sources of Cobalt, Barry M. Blechman

About the Book and Author

Bioenergy, a promising alternative for developing countries, is already a key resource (in the form of fuelwood, for example) in millions of households around the world. Third World planners are exploring new technologies and uses, including the production of biogas from wastes for household cooking, the burning of wood chips under boilers to produce electricity, and the use of alcohol distilled from sugarcane as a vehicle fuel.

Based on research sponsored by the U.S. Department of Agriculture, the Agency for International Development, and the Center for Strategic and International Studies aimed at developing detailed guidelines on the problems posed by bioenergy programs, this book describes the characteristics of a wide variety of bioenergy crops and the relationship of bioenergy to national planning. The author discusses common pitfalls in planning, such as the fragility of certain soils when native forests are cleared and replanted with exotic species. He also analyzes how general cost trends, credit problems, and the pricing of alternative fuels affect the viability of various bioenergy options. Although projects can be implemented by private entrepreneurs, communities, national and provincial government, or individual smallholders, Dr. Ramsay argues that some kind of durable national or regional commitment to bioenergy programs is vital if bioenergy is to become a successful alternative fuel and play a key role in the economic development process.

William Ramsay is the Bioenergy Project Director at the Center for Strategic and International Studies.

PUBLISHED IN COOPERATION WITH
THE CENTER FOR STRATEGIC AND INTERNATIONAL STUDIES,
GEORGETOWN UNIVERSITY

CSIS Energy Policy Series
Volume I, Number 1

Bioenergy and Economic Development

Planning for Biomass Energy Programs in the Third World

William Ramsay

First published 1985 by Westview Press

Published 2019 by Routledge
52 Vanderbilt Avenue, New York, NY 10017
2 Park Square, Milton Park, Abingdon, Oxon OX14 4RN

Routledge is an imprint of the Taylor & Francis Group, an informa business

Library of Congress Cataloging in Publication Number: 85-50337

ISBN 13: 978-0-367-00820-8 (hbk)
ISBN 13: 978-0-367-15807-1 (pbk)

Contents

Tables ix
Foreword xi
Preface xiii
Acknowledgments xv

1. Introduction 1

2. Potential Bioenergy Demand 5

Appendix 2-A: Energy Balances Explained, *Joy Dunkerley* 17
Appendix 2-B: Major Uses of Biomass Fuels, *Joy Dunkerley* 23

3. Land Use 31

Appendix 3-A: Overall Summary of Biomass Statistics for Developing Nations 39
Appendix 3-B: Collection of New Land Use Data 48

4. Bioenergy Crops 51

Appendix 4-A: Species for Humid Tropics 71
Appendix 4-B: Species for Arid and Semi-Arid Lands 79
Appendix 4-C: Species for Tropical Highlands and for Temperate Regions 82
Appendix 4-D: Leucaena 83
Appendix 4-E: Species for Special Problem Situations 85
Appendix 4-F: Genetic Engineering, *Jerry Teitelbaum* 88
Appendix 4-G: Use of Woody Residues in Costa Rica 90

5. Environmental Impacts of Bioenergy Crops 91

6. Bioenergy Conversion Technologies 103

7. Environmental Impacts and Controls for Conversion Technologies 119

Appendix 7-A: Pollutants in Household Woodsmoke 132

8. The Outlook for Costs and Financing and the Implications of Scale 135

9. Infrastructure, Marketing, and Distribution 151

10. Large-Scale Commercial and Nationalized Enterprises 165

11. Community-Managed Operations 177

Appendix 11-A: A Resource Lost: Barriers to the Participation of Women in Bioenergy Projects, *Susan Piarulli* 189

12. Miscellaneous Organizational Strategies: Governments, Donors, Associations, Smallholders, Institutions 197

13. Bioenergy and National Planning 213

14. Bioenergy on a National Scale 229

15. Planning Procedures and Suggestions 249

Bibliography 255

Index 281

Tables

2-A-I	Energy Balance Sheet for France, 1980	18
2-B-I	Final Consumption of Energy in the Residential/ Commercial/Public Sector in 1980 (percentage share by fuel)	23
2-B-II	Appliance Use in Rural Areas (percentage of households in each category)	25
2-B-III	Final Consumption of Energy in Industry in 1980 (percentage share by fuel)	27
2-B-IV	Estimated Energy Balance Sheets for the Developing Countries Showing Sector Detail of Biomass Use for 1980 and 1995 (mtoe)	28
3-A-I	Land Use, 1980 (1,000 hr.)	40
3-A-II	Areas of Natural Woody Vegetation Estimated at End of 1980 (in thousand hectares)	47
4-A-I	Selected Species Worth Considering for Mean Annual Rainfall 650–1000 mm	72
4-A-II	Selected Species Worth Considering for Mean Annual Rainfall 1000–1600 mm	74
4-A-III	Selected Species Worth Considering for Mean Annual Rainfall over 1600 mm	76
4-B-I	Selected Species Worth Considering for Mean Annual Rainfall 250–400 mm	80
4-B-II	Selected Species Worth Considering for Mean Annual Rainfall 400–650 mm	81
4-E-I	Selected Species Tolerant of Particular Soil Conditions	86

Foreword

This book represents a pioneering work. From the outset, Bill Ramsay had encouraged his colleagues to embark on an ambitious effort to explain how bioenergy projects can be effectively planned and implemented.

The investigators on this project deserve commendation not only for the insights provided into the whole area of bioenergy development but also for a clarity of exposition that will appeal to the specialist as well as the layman.

The book also deserves credit for its forthrightness in attacking common myths about bioenergy planning. Thus, while this volume supports bioenergy development, it also notes that bioenergy, as in the case of fossil fuel development, has both positive and negative environmental effects in different parts of the globe. It also notes how good projects have often failed because of institutional or social constraints that could have been overcome with better planning.

Dr. Ramsay and his colleagues emphasize the complexities of evaluating the costs and benefits of bioenergy development projects. In some cases, for example, domestically produced biofuels save foreign exchange by replacing imported oil, but, in other cases, they can cause the loss of foreign exchange either by displacing export crop earnings or by necessitating the importation of foreign capital. The book similarly demonstrates the complexities and different points of view that often arise between local projects and "national energy programs." In far too many cases, national authorities often fail to see the impact that effective bioenergy projects can have on their societies because they prefer to acquire "showpiece" projects such as large-scale hydro projects or even nuclear power reactors.

Finally, the book provides detailed recommendations to overcome the pitfalls that needlessly plague too many bioenergy programs. Given the devastating problems engendered by the energy crises in the Third World and the continued expansion of population in these areas, the book's recommendations should be given urgent attention by all those involved in the development process—donor and host governments, international institutions, and private investors. Their failure to do so will lead to a tragedy of growing proportions for much of the Third World.

Charles Ebinger, Director,
Energy and Strategic Resources Program
The Center for Strategic and
International Studies/
Georgetown University

Preface

This book was inspired by the need for an overall treatment of the bioenergy question. Much information is available on the many facets of the use of biomass for energy, but overall treatments placing bioenergy in the context of economic development have been extremely limited.

Although the recent oil crisis increased the bioenergy option remains a live one. In the long run, as supplies of fossil fuels run out, renewable sources will certainly become of crucial importance. Among the renewable fuels, the key advantage of bioenergy is that it represents solar energy stored in a naturally stable form, therefore avoiding the storage problems inherent, for example, in solar and wind energy.

The use of bioenergy has progressed since 1973, despite the uncertainties in the petroleum market. Protecting and extending world fuelwood resources has been a constant source of concern, not only for energy but also for the many other economic and environmental benefits derived from the global forest resource. The increasing use of animal wastes to replace higher cost fuelwood instead of its use as fertilizer has stimulated interest in the use of the biogas option. Some countries, notably Brazil, have mounted major programs to replace gasoline by alcohol from the fermentation of biomass, while the United States and other countries have pursued the use of alcohol blended with gasoline.

A good deal of literature has arisen out of all of this activity. Surveys of fuelwood use and household energy use, especially in rural areas, have been carried out in many countries. The alcohol and gasohol options have been extensively described, in both the popular and technical press. Programs to improve efficiencies of wood burning and charcoal burning stoves have been the subject of much documentation.

Some of the special issues involved in energy from biomass have attracted substantial attention among scientists and economists. The deforestation issue has been the subject of considerable debate. Some general issues involving alcohol fuels have been widely discussed, admittedly based often on rather shaky premises: the "energy balance" and "food versus fuel" problems.

Nevertheless, there has been a gap in the literature relating to bioenergy as a serious energy option in the context of economic development. Agencies involved in international development have been faced with a deluge of proposals and proposed programs for bioenergy projects of various kinds. These can range from community woodlot schemes to setting up pilot plants to carry out the gasification of crop wastes. Such proposals are routinely accompanied by cost data and engineering specifications. Treatment of the larger problems that are involved, however, such as the role of land use, the environment, marketing and distribution, and the general relation to economic development goals and constraints have been much less available.

This book seeks to fill part of the need for an explanation of how bioenergy projects could fit into the general scheme of economic development in a developing nation. It treats in some detail those issues that are relevant to the success of bioenergy projects and to the replication of projects as a part of larger energy programs—but which are not included in the usual cost and engineering considerations.

This book is based on research sponsored by the Agency for International Development (AID), with the cooperation of the United States Department of Agriculture (USDA). Although this is not an AID or USDA publication, and does not necessarily reflect the views of either agency, it is hoped that this book can be helpful to AID and to other assistance agencies as well as to planners involved in energy issues and in general economic development problems.

William Ramsay

Acknowledgments

This book reflects the efforts of many people. It is based on the final report for a study carried out by staff members of the Center for Strategic and International Studies/Georgetown University (CSIS) and Resources for the Future, Inc. (RFF). Several appendices are attributable to single authors. In particular, appendices 2-A and 2-B were written by Joy Dunkerley, of RFF, assisted by Caroline Bouhdili. Appendix 4-F was written by Jerry Teitelbaum of CSIS. Appendix 11-A was written by Susan Piarulli of Georgetown University and CSIS.

Furthermore, the present text reflects the influence of written contributions from numerous individuals. Specifically, appendix 3-B is based on an earlier draft by Elizabeth Davis of RFF. Chapter 7 is based on a draft by Odwaldo Bueno Netto, Jr. The original drafts of chapter 11 were written by Adela Bolet, and the original drafts of chapter 13 by Richard Kessler, both of CSIS. Part of chapter 14 is based on a paper entitled "Institutional Issues in Biomass Energy Policy: The No-Man's Land Between Agricultural, Forestry, and Energy Policies in Developing Nations" delivered jointly by William Ramsay and Jean Lipman-Blumen (ARCO Professor of Public Policy, Claremont Graduate School) at the Delhi meeting of the International Association of Energy Economists, January 4-6, 1984. The otherwise unattributed portions were written by William Ramsay, who also bears final editorial responsibility.

Charles Ebinger, Director of the Energy and Strategic Resources program at CSIS, provided overall guidance and valuable advice during the course of this work. In addition to the specific writing credits mentioned above, Elizabeth Davis contributed research for chapters 3 and 4. Julia C. Allen (now of the University of California, Santa Barbara) carried out research and wrote the key background paper on which chapter 5 is based. Richard Kessler contributed an early draft of chapter 7, and Reginald Brown of CSIS contributed material for chapter 8 and chapter 9.

Adela Bolet contributed major portions of the research and preliminary drafts for many chapters, most particularly for chapters 3, 12, and 14. Susan Piarulli performed invaluable editorial work and research for various chapters, as did Jerry Teitelbaum. The work benefited from contributions made as a consultant by Odwaldo Bueno Netto, Jr., especially in chapter 13.

Patricia Locacciato made invaluable research, editorial, and secretarial contributions to this document.

The work also benefited from important support staff contributions from Angela Blake at RFF and from, especially, Opal Burt at CSIS, as well as Jaqueline Baker, Lynda Husband, Linda Ouzts, Lenda Walker, and Pauline Younger.

Rona Fields, Mikael Grut, David Gushee, Charles Pearson, Fred Sanderson, Roger Sedjo, Deborah Shapley, Margaret Skutsch, and Fred Sussine were kind enough to provide useful comments and suggestions, many of which are reflected in the final text.

This work would not have been possible without the vision, encouragement, and support of Paul Weatherly and Alan Jacobs of the AID Office of Energy; their creative thinking in this area played an essential role in defining the scope and emphasis of the study this book is based on. The help of John Hyslop of the USDA, including an unusually helpful critical reading of the manuscript, is also gratefully acknowledged. Very useful suggestions were also received from John Kadyszewski and Jerry Storey of the Tennessee Valley Authority (TVA).

The research that made this book possible was sponsored through Cooperative Agreement Number 58-319R-4-24 between the Office of International Cooperation and Development (OICD), USDA, and Georgetown University, under the scope of work of a Participating Agency Service Agreement (PASA) (BST-4709-P-AG-3013) with the Office of Energy, Bureau of Science and Technology at AID. The book also reflects the results of earlier work carried out under the same PASA through USDA/OICD contract 53-319R-2-244 with RFF.

This book, however, does not represent an official report to AID or USDA and does not necessarily reflect the views of those agencies or any of their staff members or contractors; the conclusions are the author's alone.

1

Introduction

Bioenergy is an exceedingly promising alternative energy source for the developing countries. Bioenergy is energy derived from biomass—any sort of animal or plant tissue. Energy from biomass like trees, field crops like sugarcane, and wastes like straw and sawdust is not new; bioenergy in traditional forms like fuelwood is already the key energy source in hundreds of millions of households around the world. Planners in the Third World, however, are also interested in new technologies and new uses for this energy resource such as the production of biogas from wastes for household cooking, the burning of wood chips under boilers to produce electricity, and the use of alcohol from sugarcane as an automotive fuel.

The problem that the planner faces is apt to be not a lack of ideas and candidate projects but pervasive uncertainties about which projects are the most suitable for particular local situations and what unforeseen problems could result in any individual nation. For example, will the gasification of peanut shells help reduce the oil imports of country X? Should marginal land be used to grow trees or to grow forage crops for cattle in country Y? How should bioenergy projects like community woodlots be organized so that they work successfully and stimulate other similar projects in country Z?

This book is a guide to the bioenergy problem designed to help the bioenergy planner in a developing area deal with some of the systematic effects of introducing bioenergy projects into a specific country. The purpose of the book is not advocacy. It is assumed here that some bioenergy programs will be undertaken in the future, both in support of traditional fuel use and for the development of such modern biofuels as methane gas and alcohol from wood and other plant and animal matter. The questions treated here center on the consequences of bioenergy projects for the development process and the effects other factors in the economy of a country will have on a given bioenergy project. The emphasis is not on the engineering and cost problems that are an essential but well-studied part of the biomass energy scene. We concentrate here instead on the economic and social systems that constrain and support the individual bioenergy project.

A detailed knowledge of national energy needs is not necessary for bioenergy planning. Indeed, as stressed in chapter 2, in many countries reliable energy statistics are unavailable. Nevertheless, some feeling for the energy sources now used and for the probable demand for energy in the future is needed to avoid making costly errors. It is senseless, for example, to supply energy forms that may be irrelevant to the special circumstances surrounding the energy crisis in a particular nation.

Because land is such a critical factor in bioenergy, a detailed knowledge of land use would be ideal. In practice, however, as shown in chapter 3, bioenergy project planners selecting potential crop sites must try to make sense of the usually incomplete and conflicting data on land use that are generally available.

There are countless types of forest and fuel crops that could be utilized on candidate bioenergy project sites. As discussed in chapter 4, plantation forestry making use of species like eucalyptus and the cultivation of sugarcane to make fuel alcohol are well-established crop technologies; but more exotic species of trees, new varieties of common grain plants, and aquatic weeds such as water hyacinths may conceivably represent the shape of the future.

Although biomass as such is a ubiquitous feature of any natural environment, chapter 5 emphasizes that the planting and harvesting of biomass for energy can bring about significant environmental impacts. Characteristics of climate and soils at various sites, together with the kind of crop technology chosen, will determine the nature and extent of these effects.

Bioenergy conversion technologies presently in use include the primitive campfire, the charcoal stove, and distilleries for deriving alcohol from sugarcane. More modern processes can produce liquids and gases from the controlled heating of wood or crop wastes, methane from the fermentation of animal wastes, or electricity for modern, large-scale power plants using wood chips as boiler fuel. The technological details on these processes could fill volumes—and do. Descriptions of technologies are ordinarily available as a part of project proposals; hence chapter 6 gives only a brief review of these technologies for purposes of orientation.

All bioenergy conversion technologies will also produce environmental impacts that will have to be addressed by the planner. Effluents and controls are surveyed in chapter 7.

Cost data on proposed projects will ordinarily be available as a matter of routine. Information that the bioenergy planner will often *not* have, however, is that on the future outlook for the costs of technologies and comparisons of various scales of production, treated in chapter 8. Chapter 9 examines key problems concerning the infrastructure necessary to support a given project and investigates the status and outlook for bioenergy marketing and distribution.

A key aspect of modern bioenergy is its industrial organization. How can a project be organized to set up an efficient management structure that will provide planning and guidance and ensure long-run economic viability? Appropriate organization could take on many forms. These forms are reviewed in the next three chapters: large-scale private (or parastatal) enterprise (chapter 10), community-managed projects (chapter 11), government-run programs, associations, smallholders, and institutions (chapter 12).

A particular bioenergy project may or may not be large enough to significantly affect the national economy. Nevertheless, one pilot or demonstration project may eventually lead to an entire industry. Therefore the interaction of projects with the national economy as a whole must be taken into account, and this process is examined in chapter 13.

Chapter 14 investigates the role of the individual project in fitting into existing national bioenergy-related policy and the problems of converting individual projects into larger programs. That is, existing institutional structure and in-place public policies will be an essential background for all bioenergy development. It is in this specific local institutional context that the practical possibilities for considering the individual project as a precursor of a larger program must be considered. The success of a bioenergy program or groups of projects of significant size and scope will depend on how closely pilot and demonstration projects lend themselves to being copied and on how relevant to national planning imperatives program goals turn out to be.

Throughout the book, "planning procedure" sections at or near the end of each chapter indicate how an orderly process of bioenergy project planning can take into account each factor—demand, land use, species, etc. Chapter 15 ties these sections together into an informal planning procedure and also makes some recommendations for further research and policy actions.

2

Potential Bioenergy Demand

Introduction: Needs for New Energy Sources

Bioenergy projects will be useful in most if not all developing countries. Potential sources of supply are everywhere: some kind of plant life exists even in most of the Sahara Desert, and modern techniques of arid-land cultivation promise to make forests where vegetation is now sparse or nonexistent. Potential demand is also almost universal. Even in many oil exporting countries, traditional biomass fuels promise to retain an essential role in supplying the subsistence economy. However, in order to plan rationally for any specific bioenergy project—much less for several projects or even for an entire bioenergy program—the planner will have to have some idea of what the present energy demand pattern in a country is and how bioenergy can fit into overall energy needs.[1] For example, generating electricity by wood may not make sense for a country like Ghana where vast hydropower resources are available at reasonable cost.

We examine here some of the analytic methods involved in calculating energy demand and supply—balances and projections—and what lies behind them. We indicate how they can help the planner to identify which types of bioenergy options could fill critical gaps in the energy needs of a nation. Once these general options are known, one can go on to evaluate particular technologies for bioenergy supply and conversion—as we shall see in the following chapters.

Energy Balances

An energy balance tells what kinds of imported and domestic fuels are used (oil, hydropower, etc.) and what uses are made of them by economic sectors (industry, households, etc.). The balances usually indicate the losses of energy in uses such as electricity generation that are defined as "intermediate." They also often give estimates of the

1. At the same time, bioenergy planning cannot be expected to take on all the burdens of detailed energy or economic development data collection and analysis.

net energy (after "conversion"—usually combustion—losses) that is effectively applied to other uses. Appendix 2-A describes how energy balances are constructed and gives an example for one country.

Energy balances are not indispensable, and indeed they are often not available. Even when available, they sometimes provide only a limited amount of information. The energy balance example given for an industrial country (France) in Appendix 2-A gives a good deal of detail on industries by subsector, for example, petrochemical and food and tobacco subsectors. But many of the assessments made for developing countries during the past decade provide in fact a very modest amount of information. For example they may give only a very aggregated accounting of energy use by sectors—e.g., industry, commerce and residential, and transport.

At any rate, for the bioenergy planner it is the *future* direction of energy use that has to be examined through some form of energy demand projections. The number of automobiles in present day Sierra Leone, for example, is not the critical planning variable, but rather the number—say 50 percent higher—in the fleet 10 years hence. We see how this can be done properly in principle and, alternatively, how one can "make do" in practice.

Energy Demand (and Supply) Projections

Even in the best of circumstances, energy demand projections can only be rough estimates. Nevertheless, no planning can take place without some kind of projection, explicit or implicit. The interests of the planner are thus best served by making as explicit a projection as possible—even where it may be necessary to keep in mind data and methodology uncertainties that will severely qualify any conclusions drawn.

Economic, Demographic, and Historical Factors

Predicting the course of economic development is the name of the game when it comes to projections—but it is a difficult game to play. Usually, one must get some help from general inferences that can be made about energy trends on a worldwide basis. These general trends can sometimes be connected with the level of development and with other social and economic parameters.

Normal Population and Economic Trends and Historical Evidence. The starting point for most projections is with the present national demographic trend and with the present indicated rate of economic development. With such figures, one has some foundation for a "base case" energy use projection—assuming that one knows the connection between population, economic growth, and energy. There is, unfortunately, no magic prescription (methodology) for determining what future energy use will be even if one has a good socioeconomic crystal ball. There exist some parameters, however—specifically, the national ratios of energy per capita and energy per gross national product (GNP)—that can be helpful to the bioenergy planner. Comparative energy-economic characteristics of a number of developing countries, and especially these two parameters and how they may vary as a function of level of economic development, have been examined by analysts (see especially Dunkerley, 1980, passim).

In practice, sophisticated economic projections are very difficult to come by, and connections of economic and population growth with energy growth are also uncertain. For projecting future use patterns, therefore, energy analysts tend to rely on historical data—adjusted for unusual factors such as a threatened foreign exchange crisis. Allowances must also be made for unusual political and economic events in the past such as the effect internal strife had on Zimbabwe in the 1970s in turning a positive rate of growth into a temporary decline in GNP.

Special Economic Situations and Development Plans. In many countries, the export earnings of the economy may depend critically on commodities that are subject to large market fluctuations such as phosphates in Togo and coffee in Burundi. These fluctuations tend to depress the economy in their low phases and therefore reduce energy demand directly; in addition, the commodities then earn less foreign exchange, which could be used to buy oil. The timing and extent of the fluctuations cannot be predicted, but the records of historical market behavior can serve at a minimum to enjoin caution in projecting energy growth. In practice, past market fluctuations may at least be useful in defining high and low estimates of energy use in various sectors.

A quite common complication in the energy demand prediction problem is "planned-but-not-yet-implemented" economic development projects such as the projected opening of new copper mining complexes in Papua New Guinea. If the plans are sufficiently firm, it is possible to specify the additional electricity demand, for example, for the new facilities; but if the plans are indefinite, they will necessarily cause additional uncertainty in any projections made.

The Question of Traditional Fuels. Data on the use of traditional fuels—fuelwood, charcoal, and waste—may be deficient in quality and quantity. Some national energy planning may deal with this question by projecting as a rough estimate that present use will grow proportionally to population. Bioenergy planners must often be satisfied with such makeshift estimates. At times, however, collection of new data in the form of surveys of fuel use at the household level, for example, may be a justifiable expense. Although the number of scientific surveys already carried out is not large, there has been a good deal of investigation of problems in this area, and recommended procedures are available (FAO, 1983a).

Supply Outlook

Projections for future energy use should be modified to take into account prospects of new domestic energy supplies—not only of biomass, but of any kind. Oil exploration in the Sudan, the possible exploitation of known gas reserves in Morocco, and the development of large new hydroelectric projects in Costa Rica are all examples of the kind of changes in future energy supply that can be anticipated.

Supply changes will in general affect total energy demand in a complex way by changing quantities and prices in the energy market. In practice, energy projections usually take new sources of supply as simply replacing a given amount of imported oil by domestic supplies, like new hydropower facilities that replace imported petroleum used to generate electricity: an example of what is called "interfuel substitution."

Interfuel Substitution

The share of the total energy supply taken up by different fuels can change over time through interfuel substitution. It may seem obvious that at present levels of world oil prices petroleum products should come to be increasingly replaced by other cheaper fuels. But this is not necessarily true in detail—as past experience in India has indicated (Desai, 1981, pp.52-58)—especially when other fuels are not perfect substitutes or where prices are administered by the government. Some care should thus be taken in assuming which present fuel might be replaced by which alternative.

In the realm of the traditional fuels (fuelwood and wastes), the general tendency to replace fuelwood with kerosene in lower income urban settings may not have been totally inhibited by the recent oil crisis. Fragmentary evidence from Bangladesh, for example, suggests a strong shift to kerosene by new urban migrants (World Bank, 1982i, p. 4). In many countries, moreover, there will be shifts within certain main fuel sectors. For example, shifts from gasoline to diesel for automobile fuel may occur as a perverse response to official government pricing policies—which were designed to keep diesel prices low to encourage low-cost freight hauling and thereby stimulate agricultural and industrial development.

Constraints on Energy Projections

In addition to demographic, economic, energy supply, and fuel substitution factors there are various other phenomena that will affect future energy demand and may constrain—or stimulate—energy use growth. These are foreign exchange and external debt, pricing policies, and energy efficiency (conservation) measures.

Foreign Exchange and Debt. More than a decade after the 1972-1973 oil crisis, it should come as no surprise that the role of foreign exchange may be critical in national energy demand planning. Foreign exchange can affect energy growth directly, if a negative trade balance impedes the import of petroleum, for example. It can also affect energy use indirectly. A growing foreign exchange imbalance can make credit difficult to get and therefore restrict the import of essential capital goods for economic development projects. Less development will in general mean less energy use.

The foreign exchange or balance of payments problem touches on bioenergy directly in several ways, as we shall see in chapters 6 and 13, but it also affects projections of demand for all kinds of energy. The effect of foreign exchange imbalances on energy demand projections, however, although theoretically important, is very difficult to estimate. In practice, it usually enters a planner's considerations—if at all—by making international credit less available.

The bottom line on the balance of payments problem has to do with the external debt of a country and its credit worthiness. Again, in practice, the problem of trying to predict the future availability of credit is usually intractable. But the planner can at least be wary of including in his projections the energy needs or effects on fuel mixes of large developmental schemes for which the present financing outlook may be dim.

Pricing Policies. Government-administered prices can affect both energy demand and supply. In oil exporting countries like Indonesia, governments may still wish to follow policies of subsidizing kerosene for such praiseworthy goals as retarding the loss of native forest resources. In Brazil, pricing policies for pure-alcohol vehicles and price differentials between alcohol and gasoline fuels obviously affect the amount of petroleum imported. Realistically, however, present pricing policies should be already reflected in present demand patterns, and future pricing policies are impossible to predict. Therefore, though pricing will have important effects on the viability of bioenergy options (see chapter 8), it usually has only a nominal role in energy balance projections.

Energy Efficiency (Conservation). The pressures of the oil crisis have shown that energy efficiency can be vastly improved—but often is not.

Great improvements in vehicle efficiencies in developing country transport fleets are possible—but not always realized (Poole, 1983, pp. 25-26). The pattern of improvements in efficiencies in industry and commerce has been only marginally encouraging in Kenya, for example, but the future outlook for savings is great (Schipper et al., 1982, p. 48). Household energy efficiency (e.g., improved stoves) has been the subject of promotional campaigns in numerous countries—unfortunately with mixed results. Energy conservation (improvement in efficiency) is therefore already a real phenomenon and one that should become more significant in the future. Admittedly, the effect of efficiency changes on energy use projections in various countries of the world must be approached with some caution; nevertheless, it is reasonable to anticipate that conservation trends already observed elsewhere in the world will spread to all nations undergoing the development process.

Making Alternative Projections

Often a number of alternative projections for energy use can either be constructed or may be already available. In any case, it is usual practice to take one base case projection as an initial guideline. This base case would typically assume that historic trends in energy consumption will continue into the future, if that is consistent with the general outlook for the economy; it would necessarily omit unforseeable situations like a new oil crisis. From the vantage point of this base case, one can then examine alternative assumptions about other factors—for example, the development of new domestic energy resources—and trace their effects on fuel usage patterns and end use demands. In particular, planned development projects of a significant size, and especially energy-related projects, would be considered in alternate scenarios that assume that some or all of these projects are implemented either on schedule or with allowances made for "average" delays.

Making Do

We do not anticipate that the average bioenergy planner will have to carry out the energy balance and projection analyses just mentioned. The brief discussion above is primarily meant to explain the type of reasoning that will have gone into available energy balances and projections. In some nations, however, very little energy analysis of any kind will be available. For such cases, one needs to know how to make do with the available knowledge on energy use already prepared by local agencies or available in publications by international agencies (see Appendix 2-A for some ways of getting relevant international data).

Fortunately, in many countries the three prime energy sources will turn out to be petroleum, hydroelectricity, and traditional (wood and waste) fuels. Petroleum statistics are usually available from local representatives of commercial oil companies, from national oil companies like Petrobras in Brazil, or from government agencies regulating petroleum use. Electric utilities will have information on hydropower, as well as on petroleum used in generating plants.

Some information concerning usage by sectors is also usually attainable. Historical data on electricity usage and projections of future growth are also customarily available from the local electricity utilities. Sometimes key industrial firms will provide oil usage data, or petroleum product use can be traced through agencies dealing with transportation.

Reliable statistics on traditional fuels, on the other hand, are usually nonexistent. But even the most detailed energy assessments are often unable to find reliable data in this area. Wood or waste fuels are often traded outside of the cash market, and there have been very few studies of the cash markets that do exist; studies on India and the Philippines provide some of the few examples. The standard way out of this difficulty is to assume constant per capita use of traditional fuels—probably the least unsatisfactory alternative available without actually gathering household survey data.

Energy in Gaillardia: An Example of Making Do

The kind of reasoning that goes into making approximate projections can be clarified through an example. The following case of the hypothetical country "Gaillardia" illustrates the type of assumptions that may have to be made in cases where data are very poor. It should also give some guidance for handling more favorable cases.

Although Gaillardia is completely fictional, the various data problems described will correspond to similar problems in small developing countries throughout the world.

Commercial Fuels: Demography and Economics

Because there has been only one census in Gaillardia, the rate of population increase has been taken from an estimate made at a recent conference of donors—4 percent per year, with a rural population growth rate of 3 percent.

As far as the economy is concerned, we make the following assumptions about imports and exports. Most major imports are directly related to the size of the population, including grains, dairy products, electrical appliances, etc.—but excluding petroleum products, a special case. Because there is a small multiplier effect associated with general economic development, we therefore assume that for a 4 percent population growth the increase in (nonpetroleum) imports will be about 5 percent. For exports, the experience of the past few years shows stagnancy, and only 1 percent growth in merchandise exports is assumed until new economic development projects have taken hold, five years or so from now. Similarly, based on historical trends, services and nonrequited transfers (e.g., foreign aid) are expected to increase no more than 2 percent per year.

For the gross domestic product (GDP), the slow growth rates in the agricultural and manufacturing sectors in the past are assumed to continue, leaving a total growth of GDP of about 3 percent.

Commercial Fuels: The Special Case of Petroleum

Petroleum products are a special case in Gaillardia as in many countries. The rate of growth of gasoline consumption, recently 5 percent, is assumed to fall to 3 percent because of the recent removal of a high tax on diesel-burning cars. Kerosene consumption is assumed to rise by 4 percent, combining a shift from fuelwood to kerosene of 1 percent per year with the projected rural population growth rate of 3 percent. Growth in diesel use is complex because it is used for rail, ships, in electrical generation, as well as in vehicles, but is estimated at a rate of about 5 percent. Fuel oil is used in the small Gaillardian industrial sector but the largest part of demand is for electrical generation. The local utility estimates that electricity generation will grow at a rate of 7.5 percent. The new 50 megawatt (MW) plant coming online within the next few years will, however, be somewhat more energy-efficient than existing facilities. A rate of increase in fuel oil use is therefore estimated at 6 percent, assuming both efficiency changes and taking the growth in industrial use as lower than in the electrical sector. Other petroleum fuels, such as such as liquid petroleum gas (LPG) and jet fuel are assumed to follow historical trends.

Wood Fuel Consumption

Although the Gaillardian rural population is growing at about 3 percent, there has been some observed switching from fuelwood to kerosene. Therefore, if prices remain the same, it is assumed that fuelwood consumption will rise at about 2 percent per year over the next decade.

Alternative Scenarios: Petroleum Energy Demand

The effect of proposed development projects in Gaillardia on the energy economy can be treated through extra requirements for petroleum fuel, which are set at a 5 percent increase in petroleum imports over the base case each year after 1990. This in turn is expected to be correlated with an increase of about 15 percent in total exports of all goods and services by 1990 and a 20 percent increase in that figure by the year 2000.

Foreign Exchange Implications

Gaillardia has been fortunate in having had 95 percent of its foreign exchange deficits made up by foreign assistance transfers during 1983. It is anticipated that unrequited foreign assistance transfers will increase by 1990 to 25 percent above present levels and by another 25 percent (of the 1990 base) by the year 2000. Nevertheless, these payments will not be sufficient to close the anticipated foreign exchange gap: a tripling of imports—both petroleum and nonpetroleum—is anticipated together with only a doubling of exports of merchandise and services and a 50 percent increase in net capital transfers. Therefore, the situation should worsen, with the *net* current balance of payments deficit projected to rise to one-third of the deficit before foreign aid receipts in 1990 and half of the deficit before assistance receipts in the year 2000.

The Final Projection

The figures in the preceding paragraphs can then be used to make very approximate projections for Gaillardia of future energy demand—by fuels and in total—and the consequences for important factors like the national balance of payments. The apparently grim situation in our hypothetic Gaillardian economy is not unusual; domestic energy sources like bioenergy could prove important in solving these well-nigh universal problems.

Critical Energy Uses

As the Gaillardian example emphasizes, the bioenergy planner should be able to get by one means or the other a set of basic data and projections about energy use and its role in the economy. He should then be able to suggest which sectors have the greatest needs for which fuels. With this information in hand, he can see how particular bioenergy products fit into the picture. The planning priorities for the modern sector will be distinct from those for the traditional sector.

The Modern Sector—Priority Areas

A look at the national energy balance and projections or whatever substitute is available plus a rough categorization of the balance of payments and debt situation ("viable", "catastrophic") in many cases may give a good idea of where the most important priorities are. In many cases, the major fuel problem may be with petroleum, as might be expected. But it may not be obvious which sector has, or which petroleum products have, the greatest need (or potential) for replacement of petroleum by other fuels—specifically, by bioenergy. A common target will be petroleum fuels for transport, because efficient highway transport is necessary for the successful development of both agriculture and industry and there are generally few good substitutes for petroleum products like gasoline and diesel. The *mode* of transport, however, could be changed away from highways in countries that have excess rail or water transport. But of course such "modal switching" would be relevant only to the extent that these other modes require a smaller quantity of petroleum products to move the same goods the same distance.

Petroleum is also used widely to generate electricity, and the electricity sector is often a key to the expansion of motive power in industrial processing and water pumping in agriculture. On the other hand, residual oil burned to generate electricity can in many countries be replaced by new conventional hydropower stations. In such cases transport fuels could still be the number one priority for bioenergy fuel replacement.

Despite the general tendency to see gasoline and diesel replacement in transport as *the* problem, fuel problems have to be considered on a country-by-country basis. For many years one of the most important development objectives in India has been to increase the use of ground water for irrigation so as to make possible both double cropping and the cultivation of more valuable crops. In such situations, a fuel for motors to drive water pumps could be the prime national priority. In other nations, kerosene for household cooking and lighting could be in especially tight supply and inexpensive alternatives could be lacking; bioenergy for household uses could then take on added importance.

The Traditional Sector—Stabilization and Improvement

Many countries such as Haiti and Nepal have experienced severe crises in traditional fuels involving a loss of forest resources. In such situations, replanting of trees, or at least the replacement of fuelwood by other fuels, may take first priority. Traditional fuels, for rural industries and commerce as well as for household use, may also constitute the prime energy problem in some of the poorer oil-exporting countries: petroleum may be readily available in the commercial markets, but even at subsidized prices it will not be affordable in many poor households. (See Appendix 2-B for a description of present uses of traditional fuels.) In such cases, candidate projects for improving the supply of wood or charcoal, or increasing the efficiency of burning these fuels, may be of utmost importance.

Planning Procedure: Energy Demand and Its Relation to Biomass Options

The critical energy end use needs of a country must be matched to possible biomass energy options. These options include (1) conversion methods for turning biomass either directly into energy as an end use or into improved intermediate biofuels and (2) cultivation or collection schemes for increasing or improving the supply of biomass energy resources.

Bioenergy, through a number of current conversion options, can supply any type of fuel needed in the modern sector as well as the subsistence-sector end uses already fueled by wood or wastes. Irrigation pumps, for example, can be fueled by alcohol, by biogas, producer gas, or by steam engines fueled by wood or charcoal. The matching of these conversion technologies to end uses in automobiles, space heating, electricity generation, and steel mills can be carried out as described in chapter 6.

Each bioenergy option will need a source of supply—although some of these sources may be biomass already existing as wastes or biomass crops diverted from present food, feed, or fiber uses (see chapter 13 for a discussion of this problem). Biomass resources could also in principle be taken from natural forests; however, most energy planners must consider long-run sustainability as part of the national interest and would therefore tend not to favor "mining" natural forests on a nonrenewable basis.

In most cases, therefore, the biomass conversion option will require its own supply project or at the least will have to be coordinated with supply schemes. Whether this scheme involves forests or field crops, or even wastes, the planner will then be dealing with the land use and cropping questions discussed in chapter 3 and chapter 4.

Conclusion

Bioenergy project planners will not expect to be burdened with the problem of energy planning for the whole nation. Nevertheless, they typically will have to seek out and understand the data available from existing energy planning in order to see how candidate bioenergy projects might fit into national energy end use needs. These energy demand data will also help design new types of bioenergy projects that will better reflect national energy priorities. If adequate energy balances and projections and other relevant analyses are not available, the planner must be able to improvise from what data sources do exist. In any case, a basic understanding of the end use problem is essential if bioenergy projects are not to constitute mere speculative demonstration programs of questionable value in the larger national energy picture.

APPENDIX 2-A

Energy Balances Explained

by Joy Dunkerley

Energy Balance Construction

Energy balances are composed of three parts describing in turn the total energy requirements (TER) of the country, the energy transformation sector, and total final consumption.

The upper section of the balance sheet for France (see Table 2-A-I) presents energy consumption, that is, TER, by source of energy. These are the amounts of energy available to the country before any transformation (e.g., conversion to electricity) is undertaken and are arrived at by adding the production of primary sources of energy, imports of primary and secondary (e.g., thermal electricity produced from other fuels) energy, and quantities of energy in international marine bunkers (quantities delivered to all seagoing ships, including warships and fishing vessels) and subtracting from that amount exports of primary and secondary energy. Stock changes are either added or subtracted depending on whether they have been drawn down or replenished. Note that production of secondary energy such as production of petroleum products (either from imported or domestic crude) does not enter into the balance at this point; the only petroleum products included at this point are those that have been imported (12.93 mtoe).[2] The only electricity included is primary electricity (from hydropower or nuclear power) or imports; including indigenously produced petroleum products and thermally generated electricity at this stage, along with the primary energy from which they are produced, would involve double counting.

The energy balance for France in Table 2-A-I shows that TER are 198 million tons of oil equivalent (mtoe). Crude oil accounts for almost 60 percent and coal (solid fuels) for 18 percent of that amount. Nuclear energy accounts for 7.3 percent.

Although part of the energy (such as coal) included in TER can be used directly by the consumer, other energy forms such as crude oil

2. mtoe = million tons of oil equivalent. 50 mtoe = 1 million barrels of oil equivalent per day.

TABLE 2-A-I

Energy Balance Sheet for France, 1980

In million tons of oil equivalent (mtoe)

	Solid fuels	*Crude plus NGL plus feedstocks*	*Petroleum products*	*Gas*	*Nuclear power*	*Hydro plus geothermal & solar*	*Electricity*	*Total*
Indigenous production	14.64	2.35		6.33	14.35	16.56		54.23
Import	22.77	114.35	12.93	16.61			1.34	168.01
Export	−.89		−13.87	−.13			−1.08	−15.97
Int'l marine bunkers			−4.01					−4.01
Stock changes	−.82	−.61	−1.85	−.88				−4.16
TOTAL ENERGY REQUIREMENT	35.70	116.08	−6.79	21.93	14.35	16.56	.27	198.09
Returns and transfers		−.84	2.89					2.05
Statistical differences	−.01	−.44	−2.27	.00				−2.72
Gas manufacture	−.04		−.20	.00				−.24
Petroleum refineries		−114.79	108.58				−.40	−6.61
Electricity generation	−19.31		−9.25	−.97	−14.35	−16.56	22.19	−38.25
Own use and losses	−4.25			−.61			−3.82	−8.68
TOTAL FINAL CONSUMPTION	12.08		92.96	20.36			18.24	143.64
Total Industry	8.86		27.03	10.16			9.30	55.15
Iron and steel	6.26		1.30	1.57			1.31	10.45
Chemical	.66		4.53	5.99			1.81	12.98
Petrochemical		6.65						6.65
Non-ferrous metal	.00			.06			1.06	1.13

Non-metallic mineral	.01		1.15	.58	1.74
Paper and pulp	.05	1.30	.31	.38	2.03
Mining and quarrying	.05	.23	.23	.18	.68
Food and tobacco	.05	2.52	.43	.63	3.63
Wood and wood products				.14	.14
Machinery	.00	.29		.82	1.10
Transport equipment		.29		.49	.77
Construction				.07	.07
Textile and leather	.06	.29		.39	.46
Non-specified	1.51	9.94	.42	1.45	13.32
Total Transportation	.02	32.30	.00	.65	32.98
Air		2.57			2.57
Road		28.47	.00		28.48
Rail	.02	.61		.65	1.28
Internal navigation		.19			.19
Non-specified		.46			.46
Total Other	3.40	28.41	10.19	8.28	50.29
Agriculture		3.02		.13	3.15
Commerce		.23	4.39	2.24	6.86
Public services			.00		.00
Residential	3.40	23.00	5.80	5.18	37.38
Non-specified		2.16		.73	2.89
Non-energy Use					
Energy production[1]					
Non-energy production[2]		5.22			5.22

Notes: (1) Included in chemical or petrochemical Industry.
(2) Included only in Total Final Consumption.

Source: IEA, 1983, pg. 81.

must be transformed to be useful. In this process of transformation, energy losses occur so that the energy available to the consumer for consumption (TFC) is less than TER. The second section of the balance sheet (see the part enclosed between TER and TFC rows) shows how large these losses are and where they occur. The negative numbers in the last four rows of the second section of the balance sheet correspond to inputs of energy into the transformation sector while the positive numbers correspond to the amounts of secondary energy produced. The largest transformation losses are in petroleum refining and electricity generation. Practically all of the crude oil (114.79) and a small amount of electricity (0.40) is used to produce 108.58 mtoe of petroleum products. The energy lost in petroleum refineries is therefore the difference between these two, or 6.61 mtoe, which appears in the final column. Major quantities of solids (19.3 mtoe), petroleum products (9.0 mtoe), nuclear and hydro (together 30.9 mtoe) are used in generating electricity. Total inputs into this sector are 60.9 mtoe used in generating electricity. Total electricity generated is 22.19 (mtoe) leading to losses in electricity generation of 38.25 (mtoe). In the case of France losses are equal to 27 percent of total energy requirements.

There is, however, an important ambiguity in the measurement of losses sustained in electricity generation that can change the share of losses in TER. This ambiguity concerns the treatment of inputs into nonthermally generated electricity— largely hydroelectricity. In the case of thermally generated electricity, there is no problem; inputs are the oil, coal, and gas used to generate electricity expressed in tons oil equivalent or other common units. In the case of hydroelectricity, the energy equivalent of the inputs—falling water—must be imputed. There are two main conventions. The first is to assign the inputs the same energy value as the output. In this case there are no losses in hydroelectricity generation. The second is to assign to inputs into hydroelectricity the same energy value as if the electricity had been generated from oil, coal, or gas. In this case, of course, there will be substantial hypothetical losses on paper as average efficiencies of thermal generation are about 30 percent. Countries with large hydroelectricity generation will therefore have small losses if the first procedure is used and large losses if the second is used. And, more important, their total energy requirements will appear lower or higher depending on which conversion is used. In France, for example, inputs into thermal generation are assigned the value of equivalent thermal inputs, which, as we have seen, resulted in total energy requirements of 198 mtoe and losses of 27 percent. If inputs to primary electricity had been assigned the equivalent value of electricity output, total energy requirements would have been 178 mtoe and losses 7 percent of total energy requirements. The difference for Brazil, which has major hydroelectric generation, is even more striking. Depending on the conversion adopted, total energy requirements vary from 138 mtoe to 113 mtoe and losses from 25 percent of total energy requirements to 9 percent.

Each conversion has its advantages and disadvantages and both are widely used. Thus the Organization for Economic Cooperation and Development/International Energy Agency (OECD/IEA), from which our balance sheet for France is taken, uses the fossil fuel equivalent. The United Nations *Yearbook of World Energy Statistics* and The Latin American Organization for Development and Energy (OLADE) use the equivalent of electricity generated; the U.N. balance sheets give both alternatives to choose from.

The third section of the balance sheet describes the distribution of TFC by form of energy and by sector of the economy. In the case of France, it can be calculated from the table that the largest share (64.7 percent) is in the form of petroleum products while gas takes up 14.17 percent, electricity 12.7 percent, and solid fuels 8.4 percent. The main end use sectors are industry, transportation, other (which includes agriculture, commerce, public services, residential), and nonenergy use. In France, the nonenergy use sector uses only 3.6 percent of TFC. The largest share of TFC is used by the industrial (38.4 percent), residential (26 percent), and transportation (22.96 percent) sectors. In all of these sectors petroleum products are the most important form of energy while gas and electricity are about equal in importance.

Energy balances are invaluable tools in energy planning. By combining both supply and demand in a consistent framework, they provide a comprehensive picture of the energy sector and of substitution possibilities between the different forms of energy. The difficulty is that, until a few years ago, energy balances were available only for a few developing countries. In recent years, however, there has been increased activity in the area and energy balance sheets are available in published form for a wide range of countries (see UN 1983, OLADE, 1981). In addition, the World Bank has completed energy sector assessments containing energy balance material of 24 countries and plans to make assessments of 25 more. Although for internal use and not publication, some of the material may be available to planners in the countries where assessments have been made. At a minimum, energy planners in these countries will not have to start from scratch in constructing energy balances. For the few countries where there have been no previous activities, planners may have to start from the beginning. (An instructive example of such an exercise is given in Schipper et al., 1982.)

Data for Energy Balances

The amount of energy use in a nation consists in the broadest terms of the total amount produced, plus supplies imported, minus exported supplies. A customary distinction is made between "commercial" and "traditional" energy, where commercial energy includes oil, coal, gas, and electricity, and traditional energy covers fuelwood and animal and crop wastes. Basic data on commercial energy, expressed in common units of measurement such as tons of coal, or tons of oil equivalent (toe), or terajoules (10^{12} joules) are available for virtually all the countries of the world from the early 1960s to the present day in UN, 1983 (published annually). The basic entry for each country contains total consumption of energy and of the constituent fuels. Per capita energy consumption, that is, total energy consumption divided by the population, is also given. This publication also contains data for each country on production, trade, and consumption of the major forms of energy expressed in original units such as metric tons of petroleum or million kilowatt hours (kWh) of electricity. These latter tables give considerable detail on the different energy sectors but, being expressed in original units, must be converted to a common unit before being compared or added together. UN, 1983 contains a conversion factor table and lengthy notes to help in the interpretation of the tables. Much of the aggregated energy consumption data also appear in a summary but convenient reference for 69 developing countries in a recent World Bank publication (WB, 1983). The World Bank also published a highly recommended guideline (WB, 1982) with sections on common units and conversion factors and special problems encountered in collecting and using energy data.

Satisfactory data on use of biomass fuels are more difficult to obtain. A basic series covering production and imports and exports of fuelwood, charcoal, and bagasse (a residue of the sugar processing industry) for most of the countries of the world is contained in UN, 1983. These data appear in both original units and converted to tons coal equivalent. They do not include animal and crop wastes (except for bagasse) and as such would be expected to underestimate total consumption of biomass fuels. A comparison with other sources based on a more comprehensive definition of biomass fuels—notably OLADE, 1981—indicates, however, that there is no general rule. In some cases the UN, 1983 data are higher and in other cases lower. Most data in the two cases are, however, reasonably consistent.

APPENDIX 2-B

Major Uses of Biomass Fuels

by Joy Dunkerley

Introduction

Major markets for biomass fuels exist not only in the household and commercial sector but also in many countries in the industrial sector.

The Household (and Commercial) Sector

As Table 2-B-I shows, biomass is the predominant fuel in the household sector in all but two countries: Argentina and Venezuela. The

TABLE 2-B-I

Final Consumption of Energy in the Residential/Commercial/Public Sector in 1980 (percentage share by fuel)

	Biomass	Coal	Oil	Gas	Electricity
Africa					
Ivory Coast	92	—	4	—	3
Kenya	97	—	2	—	1
Malawi	99	—	0.5	—	0.5
Senegal	92	—	4	—	3
Zimbabwe	57	9	5	—	29
Asia and the South Pacific					
India	81	3	12	1	2
Indonesia	80	—	13	6	1
Korea	59	29	9	—	2
Philippines	66	—	27	—	7
Thailand	87	—	8	—	5
Latin America					
Argentina	4	—	46	32	18
Bolivia	48	—	41	—	10
Brazil	65	—	14	1	19
Chile	40	2	40	7	11
Colombia	64	3	17	—	16
Dominican Republic	79	—	14	—	7

small role of biomass in these countries is due to a combination of causes such as relatively high standards of living, high urban populations, and, as they are both oil producers, exceptionally cheap supplies of commercial fuels. Though Mexico is a relatively rich country with cheap petroleum products, biomass resources there still account for quite a high (60 percent) share of household use. This unexpectedly large share appears to be due to the higher rural population—33 percent compared with only 15 percent in Venezuela and Argentina. In developing countries where a significant part of the population still lives in rural areas, biomass provides by far the largest share of household energy, accounting for over 90 percent of total household use in the poorer energy consuming countries of Africa and Central America. The relation of rural-urban population to rural-urban biomass fuel consumption (fuelwood in particular) was recently studied in the Philippines. It was found that although 95 percent of the growers of fuelwood were in rural areas, many urban marketers owned rural fuelwood lots. The contracting sellers were almost all rural and only 12 percent of the rural population bought wood, while 67 percent of the urban population bought its wood. (Hyman, 1982a, pp. 7-8). A very significant portion of the fuelwood market is, therefore, excluded from these aggregated statistics. In some of the more prosperous countries the share of biomass fuel falls to 50-60 percent, but it is still a major share. Except for coal in a handful of countries with extensive domestic production (Zimbabwe, India, Korea), the other main household fuels are oil products (essentially kerosene) and electricity.

The major functional uses of household energy are cooking, lighting, and operation of appliances such as refrigerators, fans, irons, televisions, and radios. Some related activities such as irrigation, parboiling of rice, small-scale tobacco curing, operation of bars and small restaurants are probably also included in statistics on household uses though conceptually they may be more appropriately considered as agricultural, small industrial, or small commercial activities.[3] But cooking still accounts for by far the largest share (about 90 percent) of total household energy consumption and lighting for most of the rest.

In many rural areas cooking is done almost entirely with biomass fuels, with some small use of kerosene stoves confined to higher-

3. Although agriculture has been included in "household and other" here, it would be more appropriate, as the major economic activity in developing countries, to be given a separate category. The reason for including it with household activities is that often it cannot be separated easily from household activities and that the amounts are very small, accounting for about 2-3 percent of total energy consumption. It usually consists of electricity or diesel fuel for water pumping and irrigation. The small share of energy in agriculture is in a way misleading as it does not include, for example, the large amounts of energy used in transporting crops to market (which is included under transportation) or fertilizers, which are energy intensive products. If the energy used in fertilizers and in agricultural transport were included, the share of agriculture in total energy consumption would be much higher.

income households. There appears to be considerable variation in the type of biomass fuels used in cooking. As these fuels are often gathered by family members rather than purchased, the type of biomass used will depend on availabilities in the immediate vicinity. Subject to this qualification, and to the fact that the preparation of different foods requires different qualities of heat, there are some general preferences governing biomass fuels. Crop wastes such as rice husks, jute sticks, coconut husks, leaves, and twigs—though widely used—seem to be considered an "inferior" fuel whose consumption falls as income rises, giving way to the use of wood and charcoal. For some uses such as pottery kilns, however, these fast-burning, low-intensity feedstocks are highly desirable for more accurate temperature control. The use of animal wastes is strongly affected by cultural factors. They are widely used in India, Pakistan, and Bangladesh, but less so in other countries.

In rural areas, lighting using kerosene or electricity—where there is rural electrification—accounts for almost all of the remaining household energy use. Some electricity may, however, be used for irrigation pumps, as the "household sector" statistics may not distinguish in some countries between household and agricultural use. Appliance use in rural areas, though limited, varies considerably from country to country (see Table 2-B-II). In India, rural appliance use consists of irons, radios, fans, and some refrigeration (Samantha and Sundaram, 1983,

TABLE 2-B-II

Appliance Use in Rural Areas (percentage of households in each category)

	India		Colombia	
	Electrified	Nonelectrified	Electrified	Nonelectrified
Lighting				
Electricity	100	n.a.	97	1
Gas/kerosene	16	35	*	52
Candles	n.a.	n.a.	*	34
Flashlight	66	44	n.a.	n.a.
Stove				
Hotplate/electric	—	—	23	*
Kerosene	39	12	28	16
Wood	n.a.	n.a.	25	56
Coal	n.a.	n.a.	15	14
Leisure				
Electric radio	33	—	44	—
Transistor radio	47	39	42	65
Record player	3	—	n.a.	n.a.
Television	1	—	61	2

p. 62). Some appliances, such as charcoal irons, transistor radios, and even kerosene refrigerators, can be operated without electricity. Fans, on the other hand, one of the most popular of appliances in rural households, require electricity. Use of appliances in rural Colombia, where incomes are much higher than in India, is more extensive (Velez et al., 1983, chapter 4). In addition to higher ownership rates of those appliances used by rural households in India, there is widespread ownership of additional appliances such as television sets.

In urban areas, patterns of energy use by households are somewhat different. Cooking remains the major functional use of energy but the types of energy used are not the same. Although waste materials play some part in the cooking fuels of the very poor, almost all hosueholds depend on commercialized energy—wood, charcoal, kerosene, coal, gas, and electricity—that is traded in organized markets. Again as in rural markets, there appears to be a strong preference for different types of fuels. Thus wood, and often charcoal, are consumed mainly by the poorest households. As incomes rise consumers prefer to switch to kerosene and then to gas or electricity.[4]

In contrast to rural areas, the availability of electricity in urban areas permits greater use of electricity as opposed to kerosene lighting and more widespread ownership of electric appliances.

The Industrial Sector

The other major area where biomass is currently used is industry. As can be seen from Table 2-B-III, biomass accounts for a substantial share—a third or more—of total industrial use in some countries, notably in Africa and Central America. A large part of industrial energy use in these countries is in agricultural export industries such as sugar and coffee production. These industries could in principle become self-sufficient in energy and even provide surplus energy to other sectors if their wastes, in the form of bagasse and coffee tree residues, were efficiently used. In the past such industries were, in fact, often entirely self-sufficient in energy but declining prices of petroleum products before 1973 led to their use as substitutes for crop wastes. Since the rise in oil prices this trend has been halted but the process industries in many areas are still far from obtaining optimal efficiency in use of crop wastes. Tobacco is another agricultural industry with major energy requirements for drying. This is usually done

4. See, for example, Alam et al., 1982. In this study it was found that fuelwood consumption actually declined as incomes rose. The income elasticity of firewood consumption was −.7, which means that a 10 percent increase in income is associated with a 7 percent decline in fuelwood consumption. On the other hand, the relation between level of income and LPG consumption is highly positive. A 10 percent increase in income is associated with a 9 percent rise in LPG consumption.

TABLE 2-B-III

Final Consumption of Energy in Industry in 1980 (percentage share by fuel)

	Biomass	Coal	Oil	Gas	Electricity
Africa					
Ivory Coast	24	—	58	—	18
Kenya	30	—	63	—	7
Malawi	88	8	—	—	4
Senegal	—	—	62	—	38
Zimbabwe	3	47	4	—	46
Asia and the South Pacific					
India	6	71	9	3	11
Indonesia	8	1	80	7	3
Korea	8	30	1	9	50
Philippines	33	7	43	—	17
Thailand	14	6	57	—	23
Latin America					
Argentina	13	2	35	34	15
Bolivia	3	—	75	11	11
Brazil	29	11	41	1	17
Chile	16	14	48	4	17
Colombia	7	34	24	23	11
Dominican Republic	56	—	39	—	5

with firewood and, indeed, can be a major cause of deforestation in some areas.

Industrial use of biomass fuels is less frequent in non-processing industries. The major exception is Brazil, where as much as one-half of steel production is based on charcoal use. The reason for this untypically high share, which developed well before the rise in oil prices, is the distance of major blast furnace capacity from coal supplies on the one hand and its propinquity to ample supplies of forest resources on the other. A further reason for the continuing high share is the commitment of the steel industry—fortified by government help (see chapter 10)—to managing the forests on a sustainable basis to ensure continued supplies of charcoal at the desired level. The example of Brazil is of interest in that it demonstrates both the possibility of operating modern industries on the basis of biomass fuels and the necessary conditions for such a development. For most of the other countries, however, biomass has a relatively small, though still significant role (accounting for about 10 percent), in total industrial energy consumption.

These existing markets—household cooking and industrial use—for biomass fuels should continue to expand. Table 2-B-IV gives

TABLE 2-B-IV

Estimated Energy Balance Sheets for the Developing Countries Showing Sectoral Detail of Biomass Use for 1980 and 1995 (mtoe)

	Commercial fuels	*% of TEFR*	*Oil*	*% of TEFR*	*Other commercial fuels*	*% of TEFR*	*Biomass*	*% of TEFR*	*Grand Total Energy requirements (GTER)*	*% of GTER*
					1980					
Production	382	57	65	18	317	102	360	100	742	72
Imports (net)	288	43	295	82	−17	−2	—	—	288	28
Total energy form requirements (TEFR)	670	100	360	100	310	100	360	100	1,030	100
Losses	194	29	72	20	122	39	—	—	194	19
Total final consumption	476	71	288	80	188	61	360	100	836	81
—industry	188	28	72	20	116	37	72	20	260	25
—transport	180	27	162	45	18	6	—	—	180	17
—households, etc.	108	16	54	15	54	17	288	80	396	39
					1995					
Production	950	68	145	27	805	93	521	100	1,471	77
Imports (net)	449	32	386	73	63	7	—	—	449	23
Total energy form requirements (TERG)	1,399	100	531	100	868	100	521	100	1,920	100

Losses	490	32	53	10	437	50	26	5	516	27
Total final consumption	909	65	478	90	431	50	495	95	1,404	73
—industry	448	32	133	25	315	36	130	25	578	30
—transport	280	20	239	45	41	5	—	—	280	15
—households, etc.	182	18	106	20	76	9	365	70	547	28

Note: — means zero.
Source: World Bank, 1983 estimates of present and future energy consumption.

sectoral detail of biomass use, consistent with commercial energy use, for the oil-importing developing countries as a group. These estimates show both industrial and household markets for biomass expanding. In keeping with the changing relative position of these sectors in total energy demand as economies develop, the industrial market is expanding more rapidly than the household. In total demand terms, however, the largest market for biomass continues to be households.

3

Land Use

Introduction: Land, the Critical Resource

Bioenergy always requires land; furthermore, large amounts of land are needed to produce enough bioenergy to make any impact on the energy needs of the average nation.[1] Therefore it is essential in bioenergy planning to have an adequate knowledge of existing land use patterns. Unfortunately, existing land use data are more often than not seriously defective, so most bioenergy planners will have to learn how to make do with whatever data are available. Furthermore, they may or may not have the mandate, the resources, or the opportunity to search out better land use data. But while they may have to do their best with what they have, they must keep in touch with technological advances in land surveying and other land use study techniques that directly affect bioenergy resource planning.

Present land use data, perfect or not, are not enough: projections of future land use patterns are indispensable. For these projections, the bioenergy planner needs access to whatever technical information is available about climate, soils, watersheds, and any other factors that might be useful in locating future sites for bioenergy projects. With these data, he or she can estimate—sometimes only roughly—local land capabilities for bioenergy.[2]

Suitable bioenergy resource sites can then be identified as those areas that have the appropriate natural conditions for certain crops and also where a change in present land use is feasible. Sometimes *a priori* choices are unrestricted, and with a general knowledge about national or regional land use as a guide, broad candidate land areas for biofuel production can be identified. Choices could then be narrowed down by detailed investigations of several particularly promis-

1. The term "land use" usually refers to water as well as land, so that aquatic crops can also be included here by inference.

2. The planner may find the thought heartening that current exciting breakthroughs in plant breeding—and especially in plant tissue culture and genetic engineering—could conceivably within the next few decades make present-day estimates of land capability for crops absurdly pessimistic.

ing sites to establish their biological, social, and economic suitability. But sometimes practical considerations will dictate that only a few predetermined candidate sites will be available. Then those candidate sites can be immediately examined in light of specific land use standards to find out whether or not they are suitable for bioenergy production.

Existing Land Use

Most if not all nations will have country maps showing existing land uses or tables giving the amounts of land dedicated to various uses. Often these data, to be sure, only give information at a very high level of abstraction such as listing arable land, forest land, land in permanent crops, pasture land, forest and woodland, and "other land."[3] It is easy to see that such categories may be much too general for many bioenergy uses. For example, the vague term "other land" may include such distinct types as urban land and wastelands of various kinds: swamps, deserts, high mountain areas, eroded river valleys, etc. The level of description represented by "forest land" may also leave much to be desired. A finer classification, such as one developed by the United Nations Food and Agricultural Organization (FAO), gives significantly more planning information, dividing forest land into closed broad-leaved forest, coniferous forest, shrub formations, bamboo forest, mangrove forest, open broad-leaved forest, "fallow," and established plantations (FAO, 1981; see also Appendix 3-A below for some country examples).[4]

Other information useful to know would be what types of farming systems are actually in use on "arable land" or "cropland." One review of farming systems in the tropics (Evans, 1982, p.9) categorizes the different types of farming into shifting cultivation, settled subsistence farming, nomadic herding, livestock ranching, plantation systems (tea, rubber, etc.), and "nonagricultural, mainly desert." It should be noted that, overall, 45 percent of the total number of tropical "farming systems" are shifting cultivation, while nomadic herding takes up another 14 percent. This means that about three-fifths of all tropical farming systems depend on a complex long-standing human symbiosis with natural vegetation—a fact of life and land use that could be of basic importance for biomass planning.

Data are never sufficiently complete, and data are also expensive to collect. Bioenergy planning by itself may not always justify large land

---

3. See appendix 3-A for an example using FAO statistics in this kind of format.

4. Such vegetation maps that purport to show natural vegetation as a function of location must be used with care. Mueller-Dombois (1981, p. 125) has pointed out that vegetation maps may outline growth potential or "potential natural vegeta tion" rather than existing vegetation.

use data collection efforts on a regional or countrywide basis. But planners must be aware of the available data collection options, if only for future local use in the detailed evaluation of specific sites. Furthermore, useful data may be generated by independent sources: bioenergy planners can track new land use data collection efforts by nonenergy agencies or scientific institutions at the regional or national level.[5]

Gathering land use data nowadays tends to be a multi-tiered process. As discussed in more detail in Appendix 3-B below, the logical progression seems to be: 1) LANDSAT (satellite) imagery; 2) aerial photography and "side-looking radar"; and 3) ground surveys.[6] A combined approach progressing from small to medium to large-scale mapping is needed. Despite the fact that satellite images are unsurpassed as a starting point in scientific land classification schemes (Mueller-Dombois, 1981 p. 17-18), conventional satellite imagery alone cannot determine large-scale land capability. By the same token, aerial photography has its limitations, but it can help identify gross vegetation, topography, and land use patterns. In fact one may have to lean very heavily on aerial photography in rough or heavily vegetated areas, or where roads are lacking or transportation is otherwise very difficult. But when it is geographically (and financially) feasible, thorough ground surveys will be essential in filling in crucial details on area land use maps.

It should be emphasized that the land use data collection situation is fluid. Computer enhancement of satellite imagery specialized to the wavelengths corresponding to particular biomass resource biomes (environments) has shown promising results for investigations of guayule in Mexico, for example (Ridd and Neavez-Camacho, 1983). With such techniques, low-cost accurate satellite bioenergy data collection may become achievable.

Knowing what present land use is serves two functions for the bioenergy planner. First, it is the key to determining the existing types and quantity of vegetation. It therefore enables the planner to assess the potential production of fuelwood, wastes, and any other compo-

5. Outside funding is often available for such efforts. Land evaluation is often carried out as a form of aid, for example, by the Land Resources Division of the United Kingdom Overseas Development Corporation, by AID (through the U.S. Forest Service and Soil Conservation Service), or by the United Nations Development Program (Evans, 1982, p. 74). The UNDP, for example, has conducted tropical forest surveys (inventories) covering 90,000 square kilometers since 1960 (Nelson, 1973, p. 177). The FAO has given technical assistance for carrying out forest inventories, for instance, in the Philippines (FPL, 1980, p. 80). Land use studies have also been funded by AID, for instance, in the Dominican Republic where project "Siedra," sponsored by AID in cooperation with the USDA and Michigan State University, surveyed national biomass resources and assessed climate and soils for a biomass classification scheme. Results utilized OAS maps for soils and life zones, LANDSAT imagery, and information on present land use (FPL, 1980, p. 45).

6. See, for example, Evans, 1983, pp. 74ff for a brief discussion.

nents of current national biomass supply. Second, it provides a base data set for identifying potential land that could be devoted to bioenergy production. This potential will be discussed in the next subsection.

In practice, it may be difficult to make reliable estimates of the current amount of available biomass. Often only very general vegetation data can be obtained. Nevertheless, sometimes rough order-of-magnitude calculations are far better than nothing.[7]

Potential Land Uses

Planning for future land use builds on two sets of data: present land uses and scientific descriptions of the physical and climatic characteristics of the land. For investigating future bioenergy-related land uses, the bioenergy planner would benefit greatly from detailed scientific maps on topography, climate, soils, vegetation, and geology (Carpenter, 1981, p. 20-24). As with the general land use data problem just discussed, such maps may or may not be at hand.[8] Outside financing, however, is often available for carrying out this kind of mapping of land resources as an integral part of development planning.[9] Such physical and biological data are essential in making the best possible assessment of the potential biological productivity of the land, in preparation for any kind of bioenergy project.

One important factor is climate. Systems of classification have been developed for grouping areas into similar climatic zones—and some experts believe that working with these zones provides the best indication as to whether or not exotic species will thrive (Evans, 1982, pp. 60-64). But topography is also important, since in steep or rocky areas some crops may be feasible (e.g., trees) and others not (e.g., sugarcane). Also, the nature of local soils will determine if crops will be able to root easily, whether there is an untapped supply of nutrients in the ground, and whether or not the ground moisture and the whole moisture cycle are suitable for given bioenergy crops.[10]

Finally, the existing vegetation could constitute an opportunity, a hindrance, or an indicator. If local vegetation is capable of enrichment

---

7. For example, the Office of Technology Assessment (1980, Vol II, p. 15) made an inventory of potential biomass yield from U.S. forests by multiplying noncommercial acres of forests (205 million) by an average of 10 cubic feet growth per acre-year times 1.5 (the 1.5 corrects for the amount of above ground biomass being greater than the amount of commercial timber), getting a value of 3.7 billion dry tons annually (57 quads). Adding commercial forest acreage and its standing biomass increment yielded 27 billion tons total (430 quads).

8. Topographic maps identifying surface patterns and altitudes are often available from planning departments. Climate maps will typically be drawn from data gathered from meteorological agencies; rainfall and temperature data are especially needed. Soil maps are often available from the departments of agriculture, and can be expected to include data on both the physical and chemical properties of the soil. The vegetation maps should if possible indicate the growth potential of existing vegetation; the geological maps should show land forms in sufficient detail for identifying structural hindrances to possible new crops in the area (Carpenter, 1981, pp. 19-24).

9. See footnote 6, above.

10. See Choong, 1981, passim.

on an economic basis, it supplies an interesting opportunity for biomass cultivation. If not, local vegetation may be expensive and difficult to remove—a hindrance. As an indicator, the status of plant health and the type of local vegetation may provide valuable clues as to which kind of exotic species can be grown there.[11]

Although all the factors mentioned are important, climate—especially rain and temperature range—is perhaps the most important factor in predicting the potential for various forest or field bioenergy crops (Evans, 1982, pp. 137-139). Often not only the amount of rainfall, but also the rainfall distribution is of key importance. The length and severity of the dry season will determine whether certain species can be grown. Other species may only be practical candidates when the rainfall occurs mostly in the warmer part of the year—or alternatively in the cooler part. At any rate, in any detailed analysis, the predicted water balance is all-important.[12]

When looking for areas of maximum yield, as opposed to merely "suitable" areas, the importance of temperature and water remain; however, the relative area of ground covered by the leaves of a plant in the growing season and the average net rate of photosynthesis may take on greater importance (Benemann, 1980, pp. 2-3).

Finally, potential land uses must obviously take into account existing land use, vegetative and nonvegetative. Despite the difficulties mentioned in the previous section, much of this part of the analysis is relatively trivial and could be done on a site-by-site basis. Forseeable patterns of urban expansion and industrial development, or even predicted differences in rural population densities, however, are not always so easy to discern. Such changes will have an important impact on the planner's bioenergy land use map (Evans, 1982, p. 66).

In summary, the bioenergy planner needs the best maps that can be reasonably devised to represent the potential land use of a nation, a region, or a locality. These maps should show present land uses and possible crops that could be grown in different areas—with especially high yield areas distinguished —together with forseeable patterns of encroachment by urban development and other relevant land use trends.[13]

---

11. It can be seen that it is difficult to simplify the classification by using soil, climatic, topographic data, etc. to give one figure of merit for all crops, because demands vary from species to species (Moss, 1975, p.291).

12. The evapotranspiration, percolation, and run-off rates lead to specific water regimes that fit in with some crops and not with others. In turn, the rate of evaporation depends on the temperature, the relative humidity, and the wind speed.

13. There has been a good deal of theoretical work done on formal land classification and suitability schemes. (See for example Lee, 1957; Holdridge, 1981.)

Planning Procedure: Land Use and Potential Bioenergy Crop Sites

There are essentially two different approaches to land use planning. The first is to have a certain land use in mind and to find a suitable geographical location for that use (Hamilton, 1981, pp. 63-65). The other, opposite procedure is, given a certain land area, to determine the most suitable uses for it. In theory, a national bioenergy planner would probably prefer the first approach: given a nation or a region, find suitable locations for bioenergy supply (biomass cultivation) and conversion (e.g., gasifier) processes that would contribute crucial fuels needed in the national economy.[14] In practice, choices of land areas may be far from free, and the planner may well have to deal with the second approach of finding sound bioenergy options suited to a particular (available) geographical area.[15] In particular, land tenure practices may be a key obstacle, as discussed below, especially in chapters 11 and 13.

The first step in site selection in practice, therefore, is apt to be to look over a group of candidate sites already selected for various reasons: favorable ownership patterns, established political or economic priorities, or whatever. Then any knowledge—however small—gained from studying the potential land uses over the whole country or region (as sketched out in the previous section) may be applied to an analysis of the candidate sites to get a rough idea of their suitability. Obviously, if a number of such sites can be winnowed out as obviously unsuitable, it will be that much easier to carry out tests at the remaining sites on soils, rainfall, and other local conditions that will help in the final decision process.

Regardless of the amount of information available, some uncertainty is bound to exist.[16] Can existing scientific studies help? Unfortunately, quantitative site assessment models —relating measured tree growth to environmental factors at the site, for example—are difficult to interpret, because often one cannot convincingly separate cause from effect (Greaves and Hughes, 1976, p. 56). Even a fairly precise quantitative trial such as that carried out in the Usutu forests in Swaziland

14. This discussion of sites assumes that one is talking about forestry or field crops as a feedstock for biofuels. Actual biofuel conversion facilities such as charcoal kilns and gasifiers will have land use problems of their own. The later problems, however, will typically (but not always) be dwarfed by the tradeoffs involved in decisions on land for biomass cultivation.

15. Carpenter (1981, p. 34) recommends *not* determining the single best use of a given area, but instead determining a number of potential uses to which it is suitable and then pick the appropriate different levels of *management* that would be needed. This strategy, while based on a different set of priorities than the bioenergy planner usually has—i.e. "most barrels (of oil equivalent) for the buck"—may suggest that creative flexibility could be worth a try in difficult situations.

16. The problem in assessing sites for bioenergy is particularly difficult because large areas of land are involved, so that mistakes may exert a disproportionate influence on subsequent events.

(Evans, 1982, pp. 63-64), could be credited with only limited predictive powers.[17] But such experiments can at least identify potential limiting factors on productive cultivation.[18]

At any rate, small insights are sometimes all that can be hoped for. In desperate cases any minimal understanding of the logic behind current land use may be a life saver. As a very simple example, because most trees have lower nutritional requirements than most field crops, the success of field crops in potential site areas might suggest that silviculture would also thrive (Evans, 1982, p. 142).

In this respect, human records may be of more than a little use. The present owners and cultivators of the land should have useful experience about what grows and what does not grow in their areas. Anecdotal evidence from the men on the scene might also be invaluable in assessing the dangers from pests, diseases, and browsing animals. Finally, especially in areas where meteorological records are sparse, local memories might be consulted about large floods and severe storms of the past—phenomena that may be infrequent in occurrence, but devastating in effect.[19]

Decisions on growing bioenergy crops will have to be made on the basis of the least unsatisfactory information obtainable. At best this information will be obtainable from test trials (as emphasized in chapter 4) and from study of an adequate data base of the various factors that go into present and potential land use analyses. At worst, available indications of general climate and soil conditions—supplemented whenever possible by local knowledge—will have to serve as criteria for identifying suitable bioenergy sites.

Conclusion

Land use deals with the most important factor in the production of bioenergy, and information on land use is indispensable. Some data on land use will be found in almost all countries, but data will usually not be complete enough for a specific bioenergy purpose. There are

---

17. The study involved 61 plots of *Pinus patula* in 11 to 13 year-old stands. The top height of the trees was measured as an index of growth and numerous parameters were measured as independent variables, for example, soil "set," pH, bulk density, plot altitude, "aspect," and so on, and a regression analysis identified some relationships within a respectable degree of statistical significance.

18. The treacherous nature of predictions in this field is emphasized by detailed observations such as that by Kozlowski (1962, pp.149-157), in which he stresses the importance of carbohydrate storage in tree growth in some species. Depending on the particular species, increases in height may depend on the amount of previously stored carbohydrates; growth in trunk diameter, however, may depend directly on current photosynthesis and so be sensitive to soil moisture, light intensity, and temperature. Therefore the measurement of "growth" may be ambiguous for certain species in certain kinds of climates.

19. Note that a 1-in-20-year rain damaged only about 1 percent of the plantings in a recent project in the Philippines, but a freak typhoon in 1982 damaged 1.6 million cubic meters of wood, of which only 0.4 million cubic meters were recoverable (Hyman, 1982, p. 12-13).

many revolutionary techniques such as satellite imagery that the bioenergy planner might have contact with or may actually use. These new techniques can help with both the assessment of present land use and with projections of future land use. Future land use from the point of bioenergy is determined by climate—rain and temperature—soils, existing vegetation, and such other factors as topography and geology.

The science of quanitative assessment of potential sites for bioenergy or their development is in its infancy. Therefore, in addition to general climatic and soils data, evidence from present cultivation practices and anecdotal information on the local history of pests and severe storms and floods could be essential in the task of choosing potential bioenergy crop sites.

APPENDIX 3-A

Overall Summary of Biomass Statistics for Developing Nations

It may be illuminating for the bioenergy planner to examine some of the generalized categories of land uses of various kinds—and especially the different types of agricultural and forest lands—for various countries and for the developing areas as a whole.

For example, on the basis of somewhat older data, it appears that only about 2 percent of the land area of the humid tropics is cultivated, and an additional 10 percent is in pasture (Nelson, 1973, p. 9); on the other hand 80 percent of the total land area can be regarded as suitable for agriculture. The natural cover of the tropical areas has been estimated as savanna, 42 percent; rain forest, 30 percent; semi-deciduous and deciduous woodland, 15 percent; and desert shrubs, grasses, or nil, 13 percent (Evans, 1982, p. 7).

For an example illustrating on a country-by-country basis the types of common divisions of land used by the FAO, table 3-A-I shows a subdivision of the land for various developing countries into arable land, land under permanent crops, permanent pasture, forests and woodlands, and other land.

A further breakdown into more detailed types of vegetation is also of interest in many bioenergy planning contexts. Table 3-A-II shows FAO data on a selection of countries by vegetation type—closed and open broadleaved forests, shrubs, mangrove, bamboo, "fallow," and established plantations.

TABLE 3-A-I
Land Use, 1980
(1,000 hr.)

Country	*Total area*	*Land area*	*Arable Land*	*Permanent crops*	*Permanent Pasture*	*Forest and Woodlands*	*Other Land*
AFRICA							
Algeria	238,174	238,174	6,875	634	36,321	4,384	189,960
Angola	124,670	124,670	2,950F	550F	29,000*	53,760*	38,410
Benin	11,262	11,062	1,350F	445F	442*	3,970*	4,855
Botswana	60,037	58,537	1,360F	—	44,000F	962*	12,215
British Indian Ocean Territory	8	8	—	—	—	—	8
Burundi	2,783	2,565	1,110F	195F	910F	62	288
Cameroon	47,544	46,944	5,910F	1,020F	8,300*	25,640	6,074
Cape Verde	403	403	38*	2*	25*	1	337
Central African Republic	62,298	62,298	1,870F	75F	3,000F	39,690*	17,663
Chad	128,400	125,920	3,145F	5F	45,000*	20,500*	57,270
Comoros	217	217	95F	16F	15*	35*	76
Congo	34,200	34,150	655F	14F	10,000F	21,360*	2,121
Djibouti	2,200	2,198	1*	0	244*	6*	1,947
Egypt	100,145	99,545	2,700F	155F	—	2	96,688
Equatorial Guinea	2,805	2,805	130F	100F	104*	1,700*	771

Ethiopia	122,190	110,100	13,150F	730F	45,400F	26,750*	24,070
Gabon	26,767	25,767	290F	162F	4,700F	20,000	615
Gambia	1,130	1,000	270F	—	160F	216*	354
Ghana	23,854	32,002	1,090F	1,670F	3,470F	8,770	8,002
Guinea	24,586	24,586	1,500F	70F	3,000F	10,650*	9,366
Guinea-Bissau	3,612	2,800	255F	30*	1,280*	1,070*	165
Ivory Coast	32,246	31,800	2,740F	1,140F	3,000F	9,880*	15,040
Kenya	58,265	56,925	1,790F	485F	3,760F	2,530*	48,360
Lesotho	3,035	3,035	292	—	2,000*	—	743
Liberia	11,137	9,632	126F	245F	240*	3,760F	5,261
Libya	175,954	175,954	1,753	327	13,000	600	160,274
Madagascar	58,704	58,154	2,510F	490F	34,000*	13,470*	7,684
Malawi	11,848	9,408	2,300F	20F	1,840*	4,470*	778
Mali	124,000	122,000	2,047F	3*	30,000*	8,000*	81,950
Mauritania	103,070	103,040	192F	3F	39,250*	15,134*	48,461
Mauritius	186	185	100F	7*	7*	58*	13
Morocco	44,655	44,630	7,269F	450F	12,500*	5,195F	19,216
Mozambique	80,159	78,409	2,850F	230F	44,000*	15,460*	15,869
Namibia	82,429	82,329	655F	2F	52,906*	10,427*	18,339
Niger	126,700	126,670	3,350F	—	9,668F	3,900*	110,752
Nigeria	92,377	91,077	27,850F	2,535F	20,900F	14,900*	24,892
Reunion	251	250	51	1	9	103	86
Rwanda	2,634	2,495	720F	255F	470F	270*	780
St. Helena	31	31	2	—	2	1	26

Country	Total area	Land area	Arable Land	Permanent crops	Permanent Pasture	Forest and Woodlands	Other Land
Sao Tome and Principe	96	96	1F	35F	1F	—	59
Senegal	19,619	19,200	5,220F	5F	5,700*	5,318*	2,957
Seychelles	28	27	1*	4*	—	5	17
Sierra Leone	7,174	7,162	1,620F	146F	2,204*	2,060*	1,132
Somalia	63,766	62,734	1,050F	16F	28,850*	8,860*	23,958
Sudan	250,581	237,600	12,360F	57F	56,000F	48,940*	120,243
Swaziland	1,736	1,720	200F	4F	1,250F	100F	166
Tanzania	94,509	88,604	4,110F	1,050F	35,000F	42,138	6,306
Togo	5,679	5,439	1,355F	65F	200F	1,700*	2,119
Tunisia	16,361	4,700	3,190	1,510	2,550*	490F	7,796
Uganda	23,604	19,971	4,080F	600F	5,000*	6,060*	3,231
Upper Volta	27,420	27,380	2,550F	13F	10,000F	7,200*	7,617
Western Sahara	26,600	26,600	2*	—	5,000	—	21,598
Zaire	234,541	226,760	6,314	550F	9,221*	177,610	33,615
Zambia	75,261	74,072	5,100F	8F	35,000F	20,450	13,514
Zimbabwe	39,058	38,667	2,465F	74F	4,856*	23,810*	7,462
CENTRAL AMERICA AND THE CARIBBEAN							
Antigua	44	44	8*	—	3*	7*	26
Bahamas	1,394	1,007	2F	14F	1F	324*	666

Barbados	43	43	33F	—	4*	—	6
Belize	2,296	2,280	45	7	44	1,012*	1,172
Bermuda	5	5	—	—	—	1	4
British Virgin Islands	15	15	2F	1F	5F	1*	6
Cayman Islands	26	26	—	—	2*	6*	18
Costa Rica	5,070	5,066	283F	207F	1,558*	1,830*	1,188
Cuba	11,452	11,452	2,525F	675F	2,523*	1,900F	3,829
Dominica	75	75	7*	10*	2*	31*	25
Dominican Republic	4,873	4,838	880F	350F	1,510F	635*	1,463
El Salvador	2,104	2,072	560	165	610	140*	597
Grenada	34	34	5	9	3	3	14
Guadeloupe	178	176	38	11F	22	70	35
Guatemala	10,889	10,843	1,480F	354F	870F	4,550*	3,589
Haiti	2,775	2,756	545F	345F	508F	102*	1,256
Honduras	11,209	11,189	1,560F	197F	3,400F	4,060*	1,972
Jamaica	1,099	1,083	205F	60F	205F	305*	308
Martinique	110	106	7F	19F	25F	28	27
Mexico	197,255	192,304	21,800F	1,530F	74,499	48,500*	45,975
Montserrat	10	10	1*	—	1*	4*	4
Netherland Antilles	96	96	8F	—	—	—	88
Nicaragua	13,000	11,875	1,340F	176*	3,420F	4,480*	2,459
Panama	7,708	7,599	458F	116F	1,161F	4,170*	1,694

Country	Total area	Land area	Arable Land	Permanent crops	Permanent Pasture	Forest and Woodlands	Other Land
Puerto Rico	890	886	77*	62*	336	178	233
St. Kitts-Nevis	36	36	8*	6*	1*	6*	15
St. Lucia	62	61	5*	12*	3*	11*	30
St. Vincent	34	34	13*	4*	2*	14*	1
Trinidad and Tobago	513	513	70F	88F	11*	230*	114
Turks and Caicos Islands	43	43	1*	—	—	—	42
SOUTH AMERICA							
Argentina	276,689	273,669	25,150F	10,050F	143,200F	60,050F	35,219
Bolivia	109,858	108,439	3,250F	120	27,050F	56,200F	21,819
Brazil	851,197	845,651	53,500F	8,450F	159,000F	575,000*	49,701
Chile	75,695	74,880	5,332F	198F	11,880F	15,460*	42,010
Columbia	113,891	103,870	4,050F	1,600F	30,000F	533,000*	14,920
Ecuador	28,356	27,684	1,755F	865F	2,560F	14,550F	7,954
Falkland Islands	1,217	1,217	—	—	1,200*	—	17
French Guiana	9,100	8,915	3F	1F	6*	7,300	1,605
Guyana	21,497	19,685	365F	15F	999*	18,190*	116
Paraquay	40,675	39,730	1,620F	300F	15,600F	20,600F	1,610
Peru	128,522	128,000	3,100F	300F	27,120*	70,900*	26,580
Suriname	16,327	16,147	40F	12F	10F	15,530*	555

Uruguay	17,622	17,362	1,850F	60F	13,819	560*	1,073
Venezuela	91,205	88,205	3,080F	675F	17,200F	34,990*	32,260
ASIA							
Afghanistan	64,750	64,750	7,910F	140F	50,000F	1,900F	4,800
Bahrain	62	62	1	1	4*	—	56
Bangladesh	14,400	13,391	8,928F	217F	600F	2,196F	1,450
Bhutan	4,700	4,700	87	6	217	3,260	1,130
Brunei	577	527	4F	5F	6*	415F	97
Burma	67,655	65,774	9,573	450F	361*	32,167F	23,223
China	959,696	930,496	98,430F	770F	220,000F	116,400F	494,896
Cyprus	925	924	365	67	93	171	228
East Timor	1,493	1,493	70F	10F	150F	1,100*	163
Hong Kong	104	100	7	—	1	13	75
India	328,759	297,319	165,200F	3,930F	12,000F	67,480F	48,709
Indonesia	190,435	181,135	14,200F	5,300F	1,200F	121,800*	27,835
Iran	164,800	163,600	15,330F	620F	44,000*	18,000*	85,650
Iraq	43,492	43,397	5,250F	200F	4,000*	1,500*	32,447
Israel	2,077	2,033	325	88	818	116	686
Jordan	9,774	9,718	1,190F	190F	100F	125*	8,113
Kampuchea	18,104	17,652	2,900F	146F	580F	13,372*	654
North Korea	12,054	12,041	215F	90F	50F	8,970*	781
South Korea	9,848	9,819	2,060	136	48F	6,565F	1,010
Kuwait	1,782	1,782	1	—	134	2	1,645
Lao	23,680	23,080	860F	20F	800*	13,000*	8,400
Lebanon	1,040	1,023	240F	108F	10*	73F	592
Macau	2	2	—	—	—	—	2

Country	Total area	Land area	Arable Land	Permanent crops	Permanent Pasture	Forest and Woodlands	Other Land
Malaysia	32,975	32,855	1,000F	3,310F	27*	22,390*	6,128
Maldives	30	30	3*	—	1F	1F	25
Mongolia	156,500	156,500	1,182	—	123,405	15,178	16,735
Nepal	.14,080	13,680	2,316F	14F	1,786*	4,450F	5,114
Oman	21,246	21,246	18F	23F	1,000F	—	20,205
Pakistan	80,394	77,872	20,030F	290F	5,000F	2,800F	49,752
Philippines	30,000	9,920F	7050F	2870F	1,000F	12,300F	6,597
Qatar	1,100	1,100	2	—	50F	—	1,048
Saudi Arabia	214,969	214,969	1,040F	65F	85,000*	1,601*	127,263
Singapore	58	57	2*	6*	—	3*	46
Sri Lanka	6,561	6,474	1,025	1,122	439*	2,383	1,505
Syria	18,518	18,405	5,230	454	8,378	466	3,877
Thailand	51,400	51,177	16,250F	1,720F	308*	15,790*	17,109
Turkey	78,058	77,076	25,354	3,125	9,700*	20,199	18,698
United Arab Emirates	8,360	8,360	6	7	200F	2*	8,145
Vietnam	32,956	32,536	5,595F	460F	4,870F	10,330*	11,281
Yemen (North)	19,500	19,500	2,740F	50F	7,000F	1,600*	8,110
Yemen (South)	33,297	33,297	187F	20F	9,065F	2,450F	21,575

Source: FAO, 1981, pp.45–46
*Unofficial figure
F FAO estimate

TABLE 3-A-II
Areas of Natural Woody Vegetation Estimated at End of 1980
(in thousand hectares)

Country	*Closed Broad-Leaved Forest*	*Coniferous Forests*	*Shrub Formations*	*Bamboo Forests*	*Mangrove Forests*	*Open Broad-Leaved Forests*	*Fallow*	*Areas of Established Plantations*	*% Country under natural Woody Vegetation (including Plantations)*
Bangladesh	927	—	—	—	405	—	315	128	6.5
Bhutan	1490	610	25	—	—	40	205	7	45.9
India	46044	4357	5378	1440	96	5390	9470	2068	17.4
Nepal	1610	330	230	1	—	180	110	19	15.0
Pakistan	860	1305	1105	—	345	295	—	160	3.1
Sri Lanka	1659	—	215	—	4	—	853	112	25.3
Burma	31193	116	2600	632	812	E	18100	16	47.1
Thailand	8135	200	500	900	313	6440	800	114	23.0
Brunei	323	—	—	—	7	—	237	—	56.0
Indonesia	113575	190	23900	—	2500	3000	17360	1918	60.9
Malaysia	20995	18	—	—	674	—	4825	26	63.5
Philippines	9320	250	—	—	240	—	E	300	31.7
Kampuchea	7150	170	400	380	10	5100	200	7	69.9
Laos	7560	520	735	600	—	5215	5000	11	57.5
Viet Nam	7400		330	1200	320	1340	10750	204	30.2
Papua New Guinea	33710		85	0	553	3945	1380	22	82.7

Source: FAO, 1981a
E = Negligible

APPENDIX 3-B

Collection of New Land Use Data

Existing data on land resources may not be adequate to meet the needs of the bioenergy planner. If time and money permit, the planner ideally should try to obtain detailed national land use and land potentiality data—or at least information on the existing land uses and on the soil and climatic conditions of various candidate bioenergy sites. In general, data will not exist in enough detail and will show inconsistencies. For example, statistics on forest land such as those published by the FAO include such diverse environments as tropical rain-forests, temperate forests, savannas with a very low density of trees, and wastelands that are intended for afforestation (see Appendix 3-A). Furthermore, data on land resources collected for one purpose such as commercial forestry or urban land use surveys may not in all cases be applicable to bioenergy planning. Although inadequate land use data is a problem that is unlikely to be solved during investigations for any one candidate project, planning for a bioenergy project could, as a by-product, identify for future reference especially weak areas in land use data.

A number of methods for obtaining data on land use and land potential may be available to help the bioenergy planner improve upon existing data. LANDSAT imagery, radar images, and aerial photography may be used to assess land resources to supplement vegetation and climatic maps that are generally prepared from photographs or surveys. Except for the still unresolved problems of interpretation and translation between these tools, the logical order for land classifications—as mentioned in the text—is to use LANDSAT imagery first, then aerial photography and radar, then ground surveys to obtain increasing levels of detail (Evans, 1982, p. 74).

LANDSAT images, taken from a satellite orbiting the earth, can be used to identify major land formations and vegetation types. Visible and near-infrared light reflected off the surface of the earth are picked up by the LANDSAT sensors. The data are then provided either in computer format or as an image or "photograph" of the earth's surface (Umali, 1981, p. 143). Through continuous scanning of the surface over time, LANDSAT images can record changing trends in land use and vegetation patterns (Conitz, 1982, p. 179). LANDSAT data are available from remote sensing centers in a number of countries, including the United States, Canada, Brazil, and Italy. (Umali, 1981, p. 143). Further, the Regional Remote Sensing Facility in Nairobi, sponsored by AID, provides training in the use and interpretation of remote sensing techniques (Conitz, 1982, p. 173). Nevertheless, LANDSAT imagery interpretation for bioenergy resources has still not been utilized to its full potential.

LANDSAT data have been used in a number of developing countries in the past decade to provide information on land resources. In Thailand, for example, a comparison of LANDSAT images taken in 1973 with those taken between 1975 and 1976 showed an estimated decrease in forested area by about 2400 square kilometers (Umali, 1981, p. 144). In the Dominican Republic, a project under AID sponsorship has used Organization of American States (OAS) soil and life zone maps and LANDSAT imagery, together with a knowledge of existing land uses, to derive an evaluation of national biomass resources (FPL, 1980, p. 45).

Although standard LANDSAT photo imagery is generally a good tool for an overview of land resources, its overall usefulness has been limited by scale and weather constraints. The standard LANDSAT images supplied by National Aeronautics and Space Administration (NASA), for example, on a scale of 1:3,400,000, are not adequately detailed for many bioenergy purposes. Recent improvements in satellite imagery technology, however, have increased its capabilities. The LANDSAT 4 built by Hughes Aircraft Company combines a thematic mapper with the multispectral scanner used earlier and can provide a spatial resolution of 30 meters as compared to the 80 meters of earlier scanners (*Science*, 1983, p. 213). It is possible, moreover, to map areas on a scale of 1:200,000 by taking satellite pictures at four different wavelengths (Guellec, 1980, pp. 41-56). As mentioned in the text, computer enhancement also shows promise as an analytic tool.

Aerial photography, by no means a new technique, is widely used in conjunction with ground surveys to determine patterns of vegetation, topography, and land use. In areas that are rough, heavily vegetated, or lacking roads, photographs taken from an airplane can provide the planner with a closer look, providing the basis for more complete mapping based on ground surveys.

One of the shortcomings of both LANDSAT imagery and aerial photography is their reliance upon visible (and infrared) solar radiation, which makes mapping difficult in areas that have frequently cloudy conditions. Imaging radar systems, in contrast, can operate under almost all weather conditions because clouds are transparent to microwaves. Therefore, a microwave radar beam sent to the earth from a satellite and then reflected back can help map details about the earth's surface. Already, man-made structures can usually be identified, and work is being done to improve the interpretation of vegetation patterns from radar images. The Shuttle Imaging Radar (SIR-A), orbited in November 1981 on the space shuttle Columbia, has a resolution three times higher than that of earlier LANDSAT sensors. The combination of radar and LANDSAT images can provide extensive data on land uses and resources (Elachi, 1982, pp. 54-61).

A related method of assessing land resources from above is by using side-looking airborne radar (SLAR). This technique involves bouncing radio waves off the surface of the earth from an airplane. Surface features are then distinguished by the strength of the returning signals, which depend upon the types of surface, its slope, roughness, and moisture content (Evans, 1982, p. 72). SLAR has been used on several occassions where persistent cloud cover makes aerial photography difficult, including the Amazon basin in Brazil and parts of the isthmus of Panama (Jensen et al., 1977, pp. 91-92).

LANDSAT imagery from satellites, radar, and aerial photography can thus provide substantial data on uses and resources over a large area. At the other end of the technological spectrum, a number of techniques have been suggested for making quick estimates of biomass resources through simple sample survey procedures. Existing vegetation can be used as a indicator of productivity by measuring the amount of litter or standing biomass (Moss, 1975, p. 301). But extensive vegetation surveys can be time-consuming and elaborate, including transects, plot studies, and random sampling. Localized ground surveys, however, can provide site-specific data for evaluating potential bioenergy sites at more modest levels of effort.[20]

20. This appendix is based on an earlier draft by Elizabeth Davis.

4

Bioenergy Crops

Introduction: The "Manufacture" of Bioenergy

Energy is neither created nor destroyed, merely changed in form; bioenergy fuels represent a form of stored solar energy. The photosynthetic organs in growing plants accumulate the energy of the rays of the sun by assembling small molecules taken from the soil and the atmosphere into the large, complex molecules that make up plant tissues. Most of the mass of plant tissue is generally in the form of long strings of cellulose and related carbohydrates—compounds of carbon, hydrogen, and oxygen.

What we call "bioenergy" is then derived by disassembling these complex molecules—cellulose and other long-chain carbohydrates, starches, and sugars—back again into simpler molecules such as the gases carbon dioxide and water vapor and releasing the rest of the stored energy in forms usable as heat. This release or "conversion" of bioenergy fuels can be done in one step, as in burning wood in a stove. It can also be done in several steps such as turning corn starch into sugar by enzymes, then using other enzymes to ferment the sugar into alcohol, and finally burning the alcohol in the cylinders of an automobile engine.

The process of converting the solar energy to biomass is known as photosynthesis. This process is the same for all plants.[1] But—as we have seen in chapter 3—the earth's surface, in terms of soils, climate, and pre-existing human, animal, and plant communities, is exceedingly complex. The number of plant and animal species on the earth is large, and a good fraction of this number could conceivably be used for bioenergy cropping. The bioenergy planner, then, might typically encounter proposals for growing (or for harvesting the natural growth of) such disparate species as pine, sugar cane, mesquite, or kelp. A brief treatment of the issues involved should help provide general guidance for planning decisions.

1. Some plants are parasitic on other plants, animals are parasitic on plants, and so on, but these complications are not important in this discussion.

In this chapter, first we briefly examine some of the standard field and forest crops used as fuels or as feedstock for fuels. Then we look at some of the exotic (non-native) and less familiar species often proposed for bioenergy use. Next, because sometimes the choice of species may not be as important as the selection of fertilizer or irrigation methods, we examine the "factor substitution" problem. We then turn our attention to the derivation of bioenergy from by-products of forestry— that is, from the residues or wastes from conventional farming and forestry activities. Finally, we discuss how some kind of assessment can be made of the plausibility of producing various types of bioenergy crops—taking into account local resources and constraints of all sorts.

Standard Field and Forest Crops

We look here at natural forests as a source of wood for fuelwood and other bioenergy uses and also examine the use of familiar starch and sugar crops in making alcohol fuels.[2] Such local or familiar species have certain advantages and disadvantages compared to other, more exotic species— "exotics" are considered in the next section.

Natural Forests

Natural forests are the major source of world biomass energy at present and are second only to oil in supplying total world energy needs. Unfortunately, natural forests, even when not displaced for purposes of rational (economically and ecologically consistent) development, often suffer at the hands of harvesting practices that tend to destroy rather than preserve them. The preservation of natural forests from exploitative uses that are irrational over the long term is therefore one of the prime targets of land use planners in developing areas. In harmony with such land use goals, bioenergy planners will want to consider the option of managing the natural forestry source rather than mining it. Indeed, some experts have pointed out that the species of the natural forests have an important advantage as a source of future harvests of biofuels in that they occupy established ecological niches—and therefore have a built-in resistance to pests, diseases, and weather fluctuations (Evans, 1982, p. 127).

The rational management of natural forests—harvesting and replanting to maintain a truly renewable resource—is well established in temperate regions and in industrial countries. Some relevant Third

2. The definition of "standard" is obviously arbitrary, in that many plantations of "exotic" trees have been in operation for the last one hundred years or more, and the use of *fuel* alcohol from starches is of recent origin. The choice of terms here is intended to emphasize familiarity with certain species in energy contexts.

World experiences have been accumulated in the more tropical environments typical of many—but of course not all—developing regions. "Enrichment planting" to supplement natural regeneration has been carried out with reports of success in the Solomon Islands, for example. Some success in a partial management of natural forests has also been reported in Malaysia and on Mindanao in the Philippines (Evans, 1982, p. 23). On the other hand, results of natural forest management programs in Nigeria have been less successful, reportedly because expertise in silviculture (forest cultivation) or knowledge of ecology was lacking (Evans, 1982, pp. 21-22). On the whole, it is probably fair to say that there has been inadequate experience in the scientific management of the tropical forest and that insufficient information exists on how to carry out the sustained management of natural (tropical) forests (Meta, 1982, pp. 2-8).

One clue on how to overcome the natural forest management dilemma may perhaps lie in the observation that, from the little experience available, success is more likely with natural forests that tend to a dominance by a single species and to rapid growth in response to light (Evans, 1982, p. 22). But such criteria are precisely the criteria acknowledged for the success of *plantations* of trees. Bioenergy planners may therefore prefer plantations over natural forests because it appears that plantations are—irrespective of whether local or exotic species are grown—more productive than natural forests in terms of the amount of wood produced per hectare (Meta 1980, p. II-6).[3]

The practical point is that a number of tropical plantations all over the world have been observed to be successful, while trying to manage natural tropical forests is a questionable option at best.[4] A key question is then whether or not plantations should prefer local indigenous species over exotics. Native species have the definite advantage that their performance in local plantations is relatively predictable and, in addition, though usually a less important factor, local types of timber might be more acceptable to local wood users (Evans, 1982, p. 127). Against these advantages, we shall see later, is the evident superiority of many exotic species in gross yield of wood and easier crop management.

Conventional Energy Field Crops for Starches and Sugars

Sugar and starch crops are not as well-established as sources of biofuels as are natural forests. Nevertheless, alcohol from sugarcane

3. Natural forests are, however, usually also more valuable for maintaining interrelated local plant and animal communities, a goal that can have local commercial importance as well as a more global ecological significance.

4. The natural tropical forests, because of high temperatures and humidity, show high rates of chemical leaching and therefore constitute a fragile environment. The vulnerability of the tropical forest to disturbances is therefore somewhat akin to that of the relative fragility of arid lands in temperate regions.

has been used as an additive to gasoline fuels, or as a substitute for them, off and on for the last 50 years, especially in Brazil. And the fermentation of sugars and starches to produce alcohol for beverages (and more recently for the chemical industry) is hardly an exotic technology: its beginnings are lost in the mists of antiquity. Some of these tried and true field crops are still strong contenders as sources of biomass to produce liquid fuels. Sugarcane has been featured in the massive Brazilian renewable energy program and in new alcohol projects in a number of tropical developing countries. Sugarcane is also an efficient converter of solar energy into sugar and other forms of biomass (mostly cellulose).[5] The main disadvantage with sugarcane is that it requires relatively fertile land and adequate water. Therefore it competes directly with agriculture for food and fiber, displacing other crops that could be more valuable.[6] But it has the advantage that the residues from the sugar-making process, called "bagasse," plus the "trash" (leaves, stems, etc.) from the harvest, can be readily burned to provide energy for the cane-milling and distillation processes. Indeed, these residues can sometimes supply energy for outside purposes (especially electricity) as well.

In more temperate climates, grains such as corn and wheat have been used to produce alcohol for gasohol blends. Corn in particular is a very efficient absorber of solar energy (OTA 1980, III B p. 113). The technology of corn and wheat fermentation and distillation is well-established. But both grains suffer from the "disadvantage" that they are exceedingly valuable as food or feed crops.

In a word, then, the most familiar sources of starch and sugar for making alcohol—sugar cane (and beets) and grains such as wheat and corn—all have the advantage of being established crops with well-understood requirements for successful cultivation. Their principal disadvantage is that they are usually too good, in that they may be too valuable as food crops to justify their diversion to producing fuel alcohol.[7]

5. Yields vary widely from region to region, but 50 tonnes per hectare of cane is a useful rule-of-thumb figure (Hall et al., 1982, p. 50). As one ton of sugar can produce roughly 70 liters of ethanol, this means that one hectare would produce about 3.5 cubic meters or approximately 2.4 tons of ethanol (World Bank, 1980, p. 17).

6. Sugar itself as a foodstuff is often an alternative use that is more valuable than the use of cane to make fuel alcohol. The situation is complicated by the uncertainties in the price of oil—because gasoline determines the potential value of fuel alcohol—and the tremendous fluctuations in the sugar market. This complex issue is addressed in chapter 13.

7. Sugar beets have some of the same advantages and disadvantages as cane. Beets can be grown in climates where cane cannot, but beets are less efficient converters of solar energy, producing about three tons of sugar per acre. Sugar beets are not usually suited to marginal lands either.

Neither corn nor wheat plants are provided with adequate sources of cellulose residues, as sugar does with bagasse, and therefore they will need outside energy in the biofuel manufacturing process. Wheat also has a high gluten content, which may cause difficulties in fermentation (OTA, 1980, vol. III, part B, p. 115).

Exotic Species and Varieties

Biomass energy planners will naturally consider not only locally familiar natural forests and field crops as potential bioenergy sources, but also exotic species. In particular, types of trees and field crop plants that have been proved elsewhere to be good bioenergy feedstocks will merit examination, but so will some promising species that have not yet been used as practical energy resources anywhere in the world.

Exotic Forest Species

We have seen that, even where there exists a thriving natural forest, attempting to manage this natural resource for the expanded production of biofuels may be ill-advised on economic or ecological grounds. Therefore non-native types of trees deserve serious consideration in bioenergy planning.[8] Furthermore, in many cases planners will want to consider establishing forest plantations on land that is not presently forested, especially on wastelands and on underutilized pasture lands. In any case, the largest possible spectrum of tree species deserves investigation at the experimental station and in the field.

Advantages and Disadvantages of Exotic Forest Species. Usually, the prime advantage of using an exotic species is fast growth. Although many naturally occuring species will have harvesting cycles measured in decades, there are many species now available that can be economically harvested after they have been in the ground for only a few years.

Exotics with other useful qualities may be chosen: they may bear wood with a high density—and therefore a high heat value per unit of volume—or they may show "robustness," the observed ability, at least in some settings, to resist adverse weather, pests, and diseases (Evans, 1982, pp.128-133). Furthermore, many exotic species now being used for fuelwood plantations have the characteristic—common in the pea family —of being able to fix nitrogen from the air for the use of plants by the action of soil bacteria living naturally on their root systems. Finally, the ability of some exotics to *coppice*, or to regrow readily from the stumps after the tree has been cut down, is exceedingly useful in minimizing the long-term costs of planting.

---

8. The planting of exotics for fuelwood has a fairly long history. As early as 1866, plantings of *Dalbergia sissoo* were originally made in the Indus Valley of Pakistan. These irrigated plots were planted to supply fuel for the railroads but are now cut for timber and fuelwood (Evans, 1982, p. 31). The concept of "taungya" (trees interplanted with field crops) was tried in Kenya as early as 1910. The Indian neem tree was introduced into the Accra Plains of Ghana in the early 1900. During the 1920s, in the Paliparan area of the Philippines, *Leucaena leucocephala* was planted to replace imperata grass and is still producing a sustained yield of 20 cubic meters of wood per hectare per year (NAS, 1980, p. 11). And in the years between the turn of the century and the end of World War II, plantations were established with species of pine and eucalyptus in South Africa, teak and eucalyptus were planted in India, and extensive plantings of eucalyptus were made in Brazil to provide wood for railroads.

The range of exotic species is especially wide for bioenergy uses. There are fewer requirements for trees used for energy than those used for fiber. As opposed to trees needed for timber or pulp, woody plants used for biofuel may have either short or crooked trunks, or wood that warps or splits as it dries. Trees with small branches may be all right, and indeed many shrubs are superb biofuel species (NAS, 1980, p. 17).[9]

The quality of *flexibility* deserves high priority when species selections are made.[10] For example, a plantation manager may chose a species with one or more of the advantages listed above, only to find out that the species does not perform as advertised in a particular local situation (Evans, 1982, pp. 128, 133). Specifically, many species that have good reputations for planting in a similar environment in another country may turn out to be particularly susceptible to local pests of one kind or another—in other words, the two environments must not have been *sufficiently* similar. Or a new exotic species may not show any specific damage or failure in cultivation, but because of local differences in temperature, humidity, or soils may show a much lower growth rate than elsewhere. Of course, sometimes the problem is that one cannot repeat the success of a species because the seeds planted may represent a subspecies (or variety or form) that is somewhat different from the subspecies planted in a successful fuelwood project elsewhere. For this reason, the "provenance" (source) of the seeds must be checked closely (Evans, 1982, pp. 130-133).

Finally, growing exotic species usually means setting up local nurseries for the growth and harvesting of seeds and producing and distributing saplings in an organized fashion. Often such operations are easier said than done; such problems will be discussed in chapters 10, 11, and 12 on bioenergy "industrial organization" problems.

Choosing Exotic Trees. The species selected must be matched to the local climate, in particular to temperatures and rainfall. A number of candidate species have already been classified under the general headings of humid tropical, arid and semi-arid tropical, and tropical highland and temperate forests.[11]

---

9. Many examples exist. *Calliandra calothyrsus*, which can be harvested on a one-to-two year rotation cycle, was planted on 34 thousand hectares of buffer zones to protect natural forests in Indonesia to replace *Leucaena*, which was considered to grow too slowly (FPL, 1980, p.60, NRC, 1983a, passim).

As another example, two species of *Lespedeza* (*L. bicolor* and *L. thunbergii*) are shrubs now grown in South Korea for fuelwood and to provide grazing for animals. They fix nitrogen and grow rapidly and have small but densely arranged branches that are considered especially suitable for rural cooking fires.

10. One example of a flexible species are two species of *pinus*, *P. caribaea and P. radiata*, which grow well outside their natural area (Evans, 1982, p. 137). *P. caribaea* is especially suited to tropical sites, while *P. radiata* is good for the subtropics and for the warmer temperate regions. Also, for upland situations a species of Eucalyptus, *E. camadulensis*, grows well over a wide range of winter rainfall, from 300 millimeters to over 2000 millimeters.

11. Tables of recommended species by climate, together with some few examples of typical species, are given in Appendices 4-A, 4-B, and 4-C. For a more detailed summary on one species of great current interest, *Leucaena leucocephala*, see Appendix 4-D.

The choice of a particular species can be a complex matter. At first sight, it might seem reassuring to note that over 85 percent of all tropical plantations are planted in just three genera of trees: *Eucalyptus*, *Pinus*, and *Tectona* (Evans, 1982, p. 40). But these choices were mostly made before the energy crisis.[12] In recent years a very large number of species have been considered as prime candidates for plantations, especially in the tropics, where the number of naturally occuring species is quite large (Evans, 1982, p. 8; NAS, 1975, passim). Furthermore, in addition to the large number of candidate species native to the tropics, a number of species native to other climates have been planted with varying success in the humid tropics (Evans, 1982, p. 14).

Significant characteristics of candidate tree species go beyond climatic compatibility alone: the place of the species in the succession of forest growth also can be important. For example, some species are characteristic of mature forests, while other species are pioneer species that tend to take the lead in the first years in natural reforestation, but play a less important part in later stages. Pioneer species are often a good choice for plantations because they tend to be rapidly growing, hardy and aggressive (competitive with other plants). Often they can withstand degraded soils and exposure to wind and drought (NAS, 1980 p. 24). They clearly include species that have the ability to survive repeated cutting, grazing, or fires. Some of these species also burn well green, avoiding the energy losses and storage problems inherent in drying wood. As always, one can have too much of a good thing, and the very characteristics that make a tree of the pioneer type desirable can become a problem. In particular, an aggressive and rapidly growing tree can well turn into an uncontrollable "weed," which under the worst circumstances could lead to the displacement of other valuable plant species and an undesirable modification of the local ecosystem.[13]

Rapid growth—whether in pioneer species or not—is a virtue in itself because wood quality is relatively unimportant for energy uses and only the total yield of biomass per unit time (number of cubic meters per year) is usually significant for most bioenergy purposes.[14] For example, two especially fast-growing varieties of eucalyptus, together with particular species of *Gmelina* and *Leucaena*, have been the prime

12. Only 18 percent of these plantations are for fuelwood, most of them being for industrial or nonfuel uses such as watershed protection, poles and posts, agro-forestry, etc. (Evans, 1982, pp. 41-42).

A table of forestry plantations by country for the years 1965 and 1980 is given in Evans (1982, p. 37).

13. In addition to being wary of weed-like aggressivity characteristics, the tendencies of some species to burn with a bad odor or to produce numerous sparks when burning should be counted against them in species selection for fuelwood—but not necessarily for industrial wood uses.

14. But volume is not everything. Projects in the Sahel were criticized for growing trees (eucalyptus) that grew twice as fast as native species—but whose light wood reportedly required twice as much supply (Hoskins, 1982, pp. 10-11).

candidates for field tests in many areas (Knowland and Ulinski, 1979 p. 18).[15] In some places even such fast-growing tree species have not been considered adequate and have been replaced by shrubs that grow even more quickly (see footnote 9).

Another factor that usually goes into the species decision matrix is the question of hardwoods versus softwoods. Advocates of hardwoods point out that many of the hardwood species coppice, provide edible leaves and bark for animal feed, and often fix nitrogen in the soil. Certain hardwoods also possess special advantages as shade trees, for shelter belts, or in some cases as producers of edible fruits (Heybroek, 1982, p. 15). Some softwoods, on the other hand, can grow successfully on more marginal lands or in less hospitable climates.

Sometimes local conditions are characterized by special environmental restrictions or weather hazards, and the hunt for species must then take into account such factors as resistance to flooding, tolerance of salt, etc. (Heybroek, 1982, p. 15). (See Appendix 4-E for examples.)

Finally, as emphasized in the preceding subsection, there is the question of flexibility or reliability. Although behavior in other similar environments can be a good guide to selection of tree species, there is really no substitute for local testing (Heybroek, 1982, p. 19). Resources are often well spent on pilot programs testing the growth patterns of particularly promising exotic species.

Energy Field Crop Species for Cellulose Production

There is no special reason why only woody plants (trees and shrubs) must be considered as sources of cellulose and therefore as bioenergy feedstocks. Field crops of many kinds are very efficient converters of solar energy into biomass. Corn and sugarcane are examples of "modified grasses" that are efficient "storehouses" of energy. Unfortunately—for energy, not for nutrition—such crops are also economically valuable as sources of food and feed. Nevertheless, under some circumstances familiar high-value grain or sugar plants or certain lesser-known varieties or related species might be economically acceptable as a source of cellulose. For example, ordinary sugarcane, if grown in unfavorable climates or on relatively poor land, may not produce economically profitable yields of sugar. But cane grown under such conditions may still produce very large amounts of cellulose. Furthermore, if one concentrates entirely on cellulose production, one can try

15. Such fast-growing species may promise average yields of greater than 20 cubic meters per hectare per year, and under the most favorable conditions as much as 50 cubic meters per hectare per year. It must be emphasized that such large yields will depend on local circumstances, and cannot be achieved everywhere. (Hall et al., 1982, p.48).

out newer varieties of sugarcane ("energy cane") that produce more cellulose and therefore more potential biomass energy than ordinary cane. This possibility has been investigated particularly as an option for Puerto Rico (Alexander, 1982, pp. 2-12) and is currently proposed for a large program in Jamaica. A related idea is to grow tropical annual sorghums (*Sorghum vulgare* subspecies) in regions where the days are too short to produce flowering. Under those circumstances, the main energy of the plant goes into producing biomass and potentially large yields (40 tons per hectare) are possible (OTA, 1980, vol. III, part B, p. 194).

There are a number of other vegetable or fiber crops that could conceivably be adapted to energy production. These include kenaf (*Hibiscus cannabinus*), Sunn hemp (*Crotalarea juncea*), okra (*Hibiscus esculentus*), and bamboo (*Phyllostachys bambusoides*). A major problem with such schemes is that optimal yields will still depend on suitable soils, adequate water, and an adequate supply of nutrients, particularly nitrogen. Nevertheless, if soils and water are available for such high-yield crops, one may be able to afford to supply nitrogen and other fertilizing elements.

The opposite difficulty—fertilizer instead of soil quality and water supply as a key constraint—comes up when one considers the use of "ordinary" grasses. Any number of grass species can be grown in situations not competitive with field crops, both in temperate and tropical regions.[16] The difficulty with grasses is that, although they may be able to grow on lands that are marginal for field crops, total yields of biomass are much lower, say, 5 tons per hectare per year. At the same time, the nitrogen demands of the grasses are rather large, on the order of 50-100 kilograms per hectare per year, or 1-2 percent by weight of the biomass yield. An ingenious way of getting around this difficulty is to interplant nitrogen-fixing crops with the grasses. In particular, it has been proposed that alfalfa can be an effective crop that will both supply feed for cattle and nitrogen for interplanted "energy grasses" (OTA, 1980, vol. III, part B, p. 195).

One further idea for avoiding the high cost (either in cash or in forgone opportunities) of lands suitable for growing grasses is to grow energy grasses as a second crop (OTA, 1980, vol. III, part B, p. 172). In temperate climates, however, double-cropping of grasses often competes against valuable second crops such as winter wheat; in many

16. Some species investigated for the United States by the Office of Technology Assessment (OTA, 1980, vol. III, part B, p. 170) are Sudan grass (*Sorghum sudanense*)—and its hybrids with *S. bicolor*—Reed canarygrass (*Phalaris arundinacea*), tall fescue (*Festuca arundinacea*), big bluestem (*Andropogan gerardi*), switchgrass (*Panicum virgatum*), Bermuda grass (*Cynodon dactylon*), Bahia grass (*Passpalum notatum*), and elephant grass (*Pennisetum purpureum*).

Species suited to more typical developing areas include the elephant grass mentioned above and *Imperata cylindrica* (Hall et al., 1982, p. 58; see also Alexander, 1982, p. 4).

areas of the tropics, double or triple cropping is also already very common. Therefore, opportunities for extra crops may be scarcer than sometimes anticipated.

As time goes on, the number of promising "new" species or varieties, especially among biofuel field crops, will probably become quite large. Now that bioenergy seems to be a viable option for many developing countries, one can expect both conventional and unconventional plant breeding techniques to produce new species that will give greater yields at lower cost, both in terms of land rent and other factor costs. Conventional plant breeding up to now has been restricted to producing plants containing larger fractions of food or fiber or more uniformity in plant shape for mechanical harvesting—characteristics that are not necessarily vital or even desirable for bioenergy uses (OTA, 1980, vol. III, part B, p. 193). But one can expect that plant breeding techniques will increasingly be applied to finding species and varieties having a greater total weight of biomass and other characteristics especially useful for bioenergy.[17]

Unconventional Energy Field Crops for Starches and Sugars

We have already seen that getting alcohol from grains and sugars is an established, practical industrial process. The only difficulty is that the costs for fuel applications are often too high.[18] Therefore planners logically should examine crops that usually do not produce commercial feedstocks for alcohol. These crops should use marginal lands unsuited for cane or grain or should be less expensive to produce in some other way.

One likely possibility for both temperate and tropical climates is sweet sorghum. Sweet sorghum has a high sugar content and also a high yield of cellulose, which can be used for the heat needed for the distilling process (OTA, 1980, vol. III, part B, p. 185). In addition, it is tolerant of a wide range of environmental conditions, it has a relatively short growing season (110-130 days), and it is more drought-tolerant than sugarcane (OTA, 1980, vol. III, part B, p. 200). In addition, its costs of production are generally cheaper than those of cane. Unfortunately, soil and water requirements for sweet sorghum may still be too stringent to make it a practical energy crop in a many areas.

---

17. The common technique called "recurrent phenotypic selection" is recommended by one source (OTA, 1980, vol. III, part B, p. 193).

See Appendix 4-F for a discussion of genetic engineering.

18. Both costs and prices may be too high: the costs of alcohol compared to gasoline are often too high, but even in those situations where they are not, the *prices* of sugar and grain as food or feedstuffs may be too high compared to that of alcohol as a gasoline substitute.

A possible answer to the problem of demanding cultivation requirements for these field crops is starchy root crops that can be grown on marginal lands. The prime example that has been proposed is cassava (manioc). In addition to tolerating bad soils and lack of fertilizers and pesticides, cassava also has the advantage that it can be harvested all year—in contrast with sugarcane—and that its yields of alcohol are potentially high (Hall et al., 1982, p. 50).[19] Cassava has one clear bioenergy disadvantage as a bioenergy feedstock in that it is a root crop and therefore produces no bagasse waste to supply fuel for process heat. It also suffers from the more subtle disadvantage that cassava has been traditionally grown only as a subsistence crop. Experiments in large-scale cultivation have encountered problems that were not anticipated such as a sensitivity to diseases. Other root crops like sweet potatoes could also be used, and some have been discussed in the context of fuel alcohol production.[20]

Vegetable Oil Crops

Rudolph Diesel gave the first exhibitions of the operation of his engine using vegetable oil as a fuel. Vegetable oil— if sufficiently refined—still works. Therefore, rising petroleum prices have naturally enough stimulated interest in the use of oils from coconuts, soybeans, and other vegetables as a substitute for diesel fuel. The technology for extracting oil is well known, the by-products of the process can usually be sold as fertilizer or animal feed, and the highest producers (coconut palms and oil palms) yield under good conditions the satisfactory amount of 1.5-5.0 tons of oil per hectare per year. Furthermore, some of these oil crops can be grown under adverse conditions: castor and sesame are drought resistant, while coconut palms will grow in infertile sandy costal areas (Hall et al., 1982, pp. 54-56). The main fly in the ointment is the same that we have already seen for the sugar crop option. At the present prices in most localities, it is much more profitable for oil-crop growers to sell their production for food consumption rather than for fuel.

---

19. For purposes of comparison, one ton of cassava root yields about 185 liters of alcohol, compared to about 70 liters of alcohol for a ton of sugarcane. At 12 tons of cassava root per hectare per year, this means that the liters yielded per hectare per year would be 2160, compared to 3500 for a sugarcane operation yielding 50 tonnes of cane per hectare per year (Hall et al., 1982, p. 53). It seems possible, however, that cassava yields might be improved as a result of future research and development to about 20 tons of hectare per year, therefore making cassava produce about the same amount of alcohol per hectare as sugarcane.

20. In addition to sweet potato (*Ipomoea batatos*), both Jerusalem artichoke (*Helianthus tuberosus*) and chicory (*Cichorium intybus*) will grow on poor soils (OTA, 1980, vol. III, part B, p. 168). Further tests on all these crops, especially for different climates, are indicated. Chicory is said to produce 12-15 tons of sugar per hectare.

Hydrocarbon-Containing Species

The most elegant solution to the oil crises is to cultivate plants that produce "botanical gasoline"—that is, short-chain hydrocarbons that can be used directly as a liquid fuel. Plants producing long-chain hydrocarbons, like rubber trees and guayule, have long been commercially cultivated. Interest in some of the many potential hydrocarbon-producing plants has recently centered on the desert shrub jojoba (*Simmondsia californica*) and the Mediterranean-climate shrub *Euphorbia lathyris*.[21]

Euphorbia has been investigated under a privately-financed process in Arizona, with results that are as yet unpublished. One known negative feature is that the cultivation of *Euphorbia* in arid regions requires irrigation. Until more experience is available, the commercial outlook for hydrocarbon plants like *Euphorbia* remains problematical.

Aquatic Species

Aquatic, or water-growing plants are a very intriguing option for growing cheap biomass energy because in most cases they do not compete with other (dry) land uses. At the outbreak of the energy crisis, interest centered on kelp. Kelp-type seaweeds have long been harvested in China and Japan for food and chemicals.[22] It is known that kelp are not finicky feeders, so they can readily grow in coastal waters that are relatively rich in phosphorus and nitrates. A test farm operating off the coast of Catalina Island, California, has examined the kelp farming option. Methane gas was generated as the primary product, along with other useful by-products of the conversion process such as carbon dioxide, sulfur, iodine, bromine, and mannitol (GRI, 1982, pp. 5.5-2–5.5-4).

Lately, there has been more interest in cultivating freshwater aquatic species. But raising freshwater species is not as straightforward as might be supposed. Like all plants, aquatic biomass requires both water and nutrients (Benemann, 1980, p. 70), and when plants are harvested for energy use, nutrients are inevitably lost to the aquatic ecosystem. Therefore, the problem is to secure free or very inexpensive nutrients or to produce very large yields that will make the adding of expensive nutrients practical. Waste water is one possibility—where it is available.

Another possibility is to reduce bioenergy expenses greatly by reducing or eliminating planting and tending costs, that is, by harvesting

---

21. Jojoba, it turns out, appears to be more valuable, if anything, as a source of chemical feedstocks than as a source of fuel.

22. The methods used in the Orient for commercial harvesting are probably too labor-intensive for energy purposes.

"weeds." In particular, an aggressive weed like water hyacinth (*Eichhornia crassipes*) is, first of all, a high producer of biomass—current estimated yields are 60 tons per hectare per year, and 120 tons is conceivable with scientifically improved future yields (Benemann, 1980, p. 12). Second, it is a plant that often must in any case be removed from waterways as a navigational hazard.[23]

Another way of solving the nutrient cost problem is to go to self-sufficient crops such as planktonic microalgae, which have a high oil and lipid content useful for bioenergy. Blue-green algae, for example, are usually not limited by supplies of nitrogen, because some fixation of nitrogen is carried out by the algae themselves. The supply of phosphorus can be a problem, but a problem that could be solved by using municipal or industrial waste water as a nutrient stream (Benemann, 1980, p. 15). Algae have high growth rates, which is good news; the bad news is that such fast growth may require daily harvesting, and the costs of frequent harvesting could turn out to be a high.[24] Finally, "emergent" swamp-dwelling aquatic plants are of interest, especially *Typha* (cattails), and *Phragmites* (reeds) (Benemann, 1980, p. 67). The range of predicted productivity for reeds is 40-80 metric tons per hectare per year. If inexpensive swampland is available, it may be cheaper to produce these swamp plants than either microalgae or floating aquatic plants such as the water hyacinth.[25]

Improving Crops: Factor Substitution

All types of improvements are possible in the raising of forest and field crops for bioenergy. For example, the usual way of starting tree plantations is to raise saplings in a nursery and then transplant them into the woodlot. For nursery seedlings, however, the disturbance and readjustment occasioned by transplanting could be a disadvantage to their survival in the wild. Direct sowing of seeds could be a practical alternative in some cases, eliminating the need for a nursery and for a good deal of planting labor and time. (Evans, 1982, p. 226). Direct seeding by air was successful with *Leucaena* on Guam after World War

23. Even though at first glance the opportunity cost of growing such aquatic plants—which are very productive because of the high ratio of leaf area to surface covered (Benemann, 1980, p. 18)—is very low, harvesting water hyacinth by mechanical means may be very expensive. The nutrient problem must also be solved, unless a natural supply of nutrients is assured. Furthermore, hyacinths have a low tolerance to low temperatures and to high salinities and so may not be suitable for all areas.

See Sudan, 1982, p. 7, for possibilities of using the high water content of water hyacinth to advantage in a biogas scheme.

24. The production rates for algae are indeed good, with blue-green algae at 25 tons per hectare per year, perhaps going to 50 in improved future systems, with the corresponding results for green algae of 40 and 80 tons per hectare per year, respectively (Benemann, 1980, p. 12).

25. In Romania, a system of harvesting has been designed for swamp plants using equipment mounted on balloon tires.

II and has been used for *Leucaena* and *Sesbania* in South East Asia, as well as for pine forests in North America and eucalyptus in Australia (NRC, 1981a, pp. 40-43; NAS, 1980, p. 24). Germination survival percentages, however, are often low for direct sowing, and weed competition and weather hazards to young seedlings pose problems. Trials in Indonesia, however, emphasize the strong dependence on site and method of site preparation (NRC, 1981a, pp. 40-43).

Plant nutrition, and especially decisions about using fertilizer, is of course a central issue. Traditionally, little importance had been given to fertilizers for forests, either because rotation periods were so long that there occurred only a relatively infrequent disturbance of the nutrient cycle or because, in a pinch, new species could be introduced that could do without added nutrients (Evans, 1982, p. 246). But for fuel use, one may need to select high-yielding species that demand high nutrient usage. Furthermore, the rate of nutrient loss may be high. Short-rotation energy cropping harvests a larger percentage of the standing wood (and therefore more nutrients) and removes it more frequently. Fertilizer may thus turn out to be a must—for example, 0.3 ton per hectare of phosphorus (plus other nutrients) scheduled every four years for leucaena in a Hawaiian scheme (Brewbaker, 1980, p. 25). Unfortunately, fertilizer tends to be rather expensive compared to the quantitative value of fuelwood. Even worse, the expense might not be worthwhile: results of some experiments on the value of fertilizing after transplanting are ambiguous: sometimes yields were significantly increased, sometimes not (Evans, 1982, pp. 257-8).

Hard choices also have to made about weeding. Manual weeding has been the most common method in the developing world, but mechanical and chemical weed control have also been tried in forest plantations (Evans, 1982, pp. 37-38). The environmental impact of these methods will be discussed in chapter 5. Tending and harvesting practices also present opportunities to improve yields, but again, at a price.[26]

For field crops, there exists much more information about factor tradeoffs—labor versus herbicides, for example. The field of agricultural crop economics has been intensively and scientifically studied. Numerous investigations have been made of the effects of replacing labor by machines, the tradeoffs between increasing yields and the

26. In one project (FAO, 1982a, p. 50) a major cause of failure before 1973 was lack of tending; the official national plan from 1977 therefore made the first Saturday in November a "Forest Tending Day."

To save costs of land clearing, a scheme called "line enrichment" has been tried in some areas (Evans, 1982, p. 227). In this option, only a "matrix" or gridwork of native (usually second growth) trees is cleared, and unusually low-cost tree crops can result if suitable exotic species are chosen for planting along the lines. For line enrichment planting, one needs species that do not need to be thinned and that are fast growing; there must be few or no browsing animals, etc.

expense of fertilizers, and various other factors for increasing productivity and cost efficiency throughout the crop cycle from planting to harvesting. Hence, standard cropping methods will quite likely already reflect the most economically sound combinations of productive factors to be used.

As always, generalizations about which factor combinations are best are difficult to make because of the importance of local conditions. But it is evident that the price that can be paid for energy crops that compete with oil priced at about $30 a barrel is generally lower than the value of the food or fiber crops that compete for the same land. High-priced inputs such as fertilizer that could be justitified for food crops will often be too expensive for producing biofuels. Therefore, production factor tradeoff equations for food and fiber crops must be skeptically examined for energy field crop projects.

Wastes

Generally speaking, using wastes to provide bioenergy is an excellent idea. The only real problem is determining when a waste is really a waste.

A ready resource in many countries is the use of wood wastes from logging or from sawmill operations. A large fraction of the tree is generally left on the forest floor as waste during logging. But if some or all of these remnants—small branches, twigs, leaves, and even roots—are used for energy purposes, the ecological balance of the local plant community may be signficantly disturbed—as discussed in chapter 5. On the other hand, using sawmill wastes—sawdust, shavings, and offcuts—is another question entirely. These wastes comprise as much as half of all the wood harvested (Hall et al., 1982, p. 36). Often a good fraction of these sawtimber wastes can be used for pulp, but inevitably a large residual fraction that is ideal for energy use will remain (Watt, 1982, p. 184). Use of these wastes is not an exotic proposition; most U.S. sawmills have converted within the last decade from generating process heat with natural gas or oil to the combustion of wood wastes. Furthermore, the country-wide potential can be enormous. It is estimated that the potential energy from wood wastes in Ghana exceeds total domestic energy consumption of all kinds (Powell, 1978, p. 119); about 20 percent of these wastes are in fact already used to make charcoal (Powell, 1978, p. 120).[27] (See Appendix 4-G for estimates of costs for wood wastes in Costa Rica).

27. The amount of wood wastes potentially available for fuel in Ghana was estimated to be between 7 and 8.7 million cubic meters, of which 1.6 million cubic meters go to charcoal.

Finally, residues are available from trees used for other purposes such as fruit trees, trees bearing other edible crops like nuts, or trees used to provide shade for crops like coffee and cacao plants.

Animal and crop wastes are generally much more complicated to plan for than sawmill wood wastes. Animal wastes are commonly already used for fuel in many countries, especially in South Asia. The main difficulty is that these wastes are also valuable for fertilizer. Indeed, the "no-win" tradeoff between animal wastes for fuel and for fertilizer is a particularly dismal fact of life in many countries. On the positive side, one of the attractions of biogas (fermentation to methane) lies in the fact that the fermentation process preserves most of the nitrogen in the wastes, so that both energy and fertilizer can be produced from the same biomass (see, for example, the review in Cecelski and Ramsay, 1979, pp. 35-37).

The options for crop wastes have some similarity to those for animal wastes. It is difficult to generalize, but it has been estimated that approximately two-thirds of the "average" crop ends up as residue (Hall et al., 1982, p. 38). For a cash crop case in Hawaii, the trash from sugar fields (the unharvested part of the cane plants) had formerly been collected and used for landfills (Castberg et al., 1981, p. 94). It proved a practical bioenergy project to grind up this trash and use it, together with the processing waste, bagasse, to provide boiler fuel for both process heat and generating electricity (Castberg et al., 1981, p. 97). On the other hand, crop wastes are often intensively used, especially in subsistence agriculture, as fertilizer or as fuel in primitive stoves. Any leftover wastes may be applied to still other uses such as building materials for adobe or brick dwellings.

No one answer can be given to this problem of which waste is waste. One study on the use of waste biomass in industrial countries estimated that an average of 60-70 percent of crop residues should be harvested for bioenergy—the other 30-40 percent "residue of the residues" were to be left in place in the fields either to impede soil erosion or because the cost of collection would be too high (OTA, 1980, vol. III, part B, 14). But in general great caution is required in *any* planning for the use of such an already highly utilized commodity such as crop wastes.[28]

Food processing wastes are an in-between case. The best known food (or alcohol) processing waste in the context of bioenergy is the bagasse from sugarcane. The use of bagasse, however, is nothing new in most developing areas and utilization rates are very high. Molasses, however, is a waste from sugar processing that is not always fully

28. Questions of economy in the hypothetical processing and transportation of waste for bioenergy must be considered, of course. Densification through bailing or stacking or compression of process wastes can decrease transportation costs or lower costs of handling and combustion (OTA, 1980, vol. III, part B, p. 10).

utilized. Although molasses is very valuable as an animal feed, in many cases no local market for molasses exist. Under these circumstances, the fact that molasses from sugar processing can produce 280 liters of alcohol per ton of molasses can be intriguing (Hall et al., 1982, p. 52).[29]

Other wastes may or may not be economic to utilize for bioenergy, depending very much on local circumstances. Possible use of banana residues to make charcoal or coffee husks or peanut shells as a feedstock for gasifiers are the kind of options that might be available (Sudan, 1982, pp. 5, 6). One obvious candidate is wood wastes, especially wood leftover after land is cleared for agriculture. Although such wood is not a patently "renewable" resource, it represents a large wastage of resources in many areas of the world.

All in all, there are many interesting possibilities for the bioenergy use of wastes, but the question of the true cost of the waste is complex, and requires careful analysis.

Planning Procedure: Species Selection

As we have seen, the first step in a bioenergy assessment is to determine whether proposed or potential bioenergy projects will satisfy an existing energy demand—or a potential energy demand that would significantly assist national economic development (or survival). Next, the planner must check existing land uses and see what the availability of existing biomass crops is. If the existing biomass resources do not suffice to fill energy needs, then projects based on plantations or new field crops must be considered. This means, first, that an appropriate species must be selected.

Species selection and related project planning steps have been discussed in this chapter. We have seen that there are various general characteristics that one can use as criteria for choosing species.[30] For most purposes, however, the first thing looked at will be *yield*—more precisely, yield of burnable dry matter. As a generalization, yields tend to be relatively high in the non-arid tropics, where high insolation (exposure to sunlight) and long or continuous growing seasons are

29. A case in point is a proposed project in the Sudan. There, molasses produced in a sugar refining plant could not be shipped economically for sale on the world market because of long distances and poor transportation to port cities and local enzootic cattle diseases prevented its sale as a feed supplement. The economics of alcohol conversion consequently appeared quite favorable (Sudan, 1982, pp. 4-5).

30. Choices of tree species have been related to general criteria in one study (Burley and Wood, 1976, pp. 104-105). They are: (1) growth—dry matter produced, volume of stem, (2) morphology—crookedness, branches, etc., (3) physiology, (4) robustness and health—resistance to pests and disease, (5) technological aspects—properties of the wood, (6) chemical nature—uptake of nutrients.

very favorable for growth.[31] But it must again be emphasized that a plethora of other conditions must be satisfied if anticipated yields are actually to be realized.

To determine real-world yields, local field trials are therefore essential. But just as general information on climate does not insure that an exotic species will perform in a new specific habitat, so too even a number of small trial plots set up at a range of possible sites will not solve all problems (Hughes and Willan, 1976, p. 12). Trial plots will give information, for instance, about competing vegetation,[32] but relatively short-term trials may not give information about phenomena that occur only rarely such as flooding and fires.[33]

Species tests are at any rate only part of the story. All kinds of technical failures can arise whether the project is a new tree or field crop planting scheme or whether it involves the adaptation of existing biomass crops to new fuel conversion technologies. Practical problems can range from nursery management snafus to import restrictions on seeds.[34] Furthermore, common sense tells us that the more complicated a project is, the more things there are that can go wrong.[35]

Technical difficulties, however, are often the least of one's worries. Sometimes local preferences, for example for a particular kind of wood species for domestic use, can be complicated by unexpected sociological problems (See Hyman, 1982a, passim). On the other hand, sometimes apparently overwhelming "human factor" problems may have obvious solutions when they are analyzed more closely by the

31. One project with *Eucalyptus deglupta* in Papua New Guinea showed after three years a wood volume of 288 cubic meters per hectare (Evans, 1982, pp. 4-5). But again, individual site conditions have to be taken into account, and such factors as seasonality and consistency of rainfall are important—the total amount of rainfall may be large in some cases but the length or the beginning of the wet season may vary in unfavorable ways (Evans, 1982, p. 7). Again it must be emphasized that assumptions about yield without local testing may be shaky. For example, in one project in the Philippines, it was assumed that yields of *Albizia falcataria* would be at least 200 cubic meters per hectare at an age of 8 years—in fact, the research division of the agency believed that 260-310 cubic meter yields were likely. As it happened the actual yields were 180-250 cubic meters, perhaps reflecting the fact that the ideal regimen set up by technical experts at the agency was not followed by all the growers (Hyman, 1982, p. 11).

32. For example, *Pinus caribaea* has been found to be resistant as a sapling to grass, but it is not resistant to more aggressive vines in the second growth from cut-over rain forests in New Guinea. On the other hand, some hardwoods, especially eucalyptus, are particularly discouraged by grass (Evans, 1982, p. 143).

33. *Eucalyptus robusta* is unusually resistant to flooding, while *Acacia auriculiformis* and many eucalyptus are resistant to fire (Evans, 1982, p. 143).

34. Some of the things that can go wrong with tree projects are listed in Evans (1982, pp. 134-5). (1) Seed supply may fail because of high demand, poor crops, or collection difficulties (2) There may be nursery problems. (3) Some species may not be suitable for plantations: they may be climax species that may not perform well as secondary plants, or there may be structural problems (for example, an overcrowding of crowns in *Octomeles sumatrana*) or chemical difficulties may arise (for example *Grevillea robusta* supresses its neighbors). Also, depending on national conditions, there may be (4) prohibitive quarantine periods for incoming plant materials or (5) a lack of sufficient foreign exchange to purchase abroad; finally (6) there is always susceptibility to disease and pests as when shot borers (*Hypsipyla robusta*) attack the tree family Meliaceae, or *Dothistroma* damage young *Pinus radiata* seedlings.

35. For example, when agroforestry schemes are considered —that is, intercropping field crops with trees—eucalyptus are in general not good because they have high moisture requirements and tend to produce toxic substances. On the other hand, leucaena tends to fertilize its surroundings and its open leaf structure is rather favorable for agroforestry (Hall et al., 1982, p. 49).

project designer. For example, since trimmings from rubber trees in plantations have traditionally provided fuelwood in parts of Malaysia, there may be an understandable ingrained local preference for planting trees that produce other valuable products, and not just fuelwood (Knowland and Ulinski, 1979, p. 18). More generally, poles for building are in great demand in many areas, and species with suitable trunks will therefore be favored.

In a word, the question of species selection and related crop planning is fully as complicated as might be imagined. A partial solution to this difficulty is to anticipate the kinds of problems that might arise at specific sites—such as the "potential bioenergy crop sites" discussed in chapter 3. Estimates of long-range hazards should be included, as well as problems that are likely to occur in the course of a pilot project of average length.

Conclusion

The choice of species for bioenergy cultivation presents a frustrating problem in planning. On the one hand, there is almost too much information available on a vast number of species of trees, terrestrial and aquatic plants, and even one-celled plants that could be of interest for bioenergy. On the other hand, conditions of climate, soils, and topology vary so much that it is difficult to predict species behavior for specific bioenergy cultivation sites.

One of the most important choices is between native and exotic species and between conventional agriculture and more unfamiliar or exotic field crops. The advantages of native tree species and conventional crops like sugarcane lies in their reliability and predictability. For tree crops, however, this predictability is more apparent than real. The problem with field crops, on the other hand, tends to be that the costs of cultivation are too high. Moving on to exotic species, there are numerous species of trees and shrubs that combine advantages like high yields and easy reproduction (coppicing). In addition, planners will have available to them species having a confusing variety of other attributes, such as resistance to drought and nitrogen fixation capabilities, that will affect their choices. Newer varieties of field crops will present good opportunities for the production of cellulose, although land fertility requirements may be high. The requirement for nutrients will constrain the growing of grass as well as the more exotic aquatic crops, while crops such as vegetable oils have built-in economic problems to solve.

Changing the pattern of use of factors of production like fertilizer is a possible solution to fitting crops to site capabilities. The use of wastes can obviate the apparent high costs involved in cultivation. But increased use of fertilizer or machinery costs money, and wastes are not always "wastes," so the opportunity cost picture is complex.

The ultimate goal is to carry out an in-country plausibility assessment of which species might fit into which sites. Although species can be chosen roughly on the basis of yields and other desirable characteristics, field trials are usually essential in the real world. In addition, the importance of sociological factors and such practical problems as import restrictions on seeds will have to be considered.

APPENDIX 4-A

Species for Humid Tropics

Some of the most important determinants of the range of a candidate tree species for bioenergy plantations are rainfall and temperature—or altitude, because temperature tends to drop with height above sea level. Tables 4-A-I, 4-A-II, and 4-A-III each give possible species for different temperature (or altitude) ranges, and for mean annual rainfalls of, respectively, 650-1000mm, 1000-1600mm, and over 1600mm.

These tables should give some idea of species selection possibilities over a wide range of climatic conditions. The two determinants here, rainfall and temperature, by no means give the whole story, however. For more detailed analyses, a number of studies are available (Webb et al., 1980; NAS, 1971, 1980, Fenton et al., 1977). (See also Appendix 4-D for a discussion of leucaena.) The range of factors can be suggested by looking at one or two cases, however.

The species *Sesbania grandiflora* has been successfully grown in plantations in Indonesia (NAS, 1980, pp. 62-63). It produces 20-25 cubic meters of wood per hectare per year, it also produces pods, flowers, and leaves used for human and animal feed, and it can be seeded from the air (NAS, 1980, pp. 24). On the other hand, the wood has a low specific gravity (0.42), which means it is relatively bulky to handle. Furthermore, there is reportedly little information in readily available literature on this species and therefore potential growers should approach it with caution (NAS, 1980, pp. 62-63).

Another popular tropical plantation species is *Gmelina arborea* (NAS, 1980, pp. 46-47). It is widely used for plantations in the humid tropics because it can be established more easily and less expensively than many other species. It regenerates well from sprouts and its silviculture is relatively well known. A tree of 20-30 meters, it produces a relatively light wood but with a good yield, probably exceeding 30 cubic meters per hectare; it produces good timber and satisfactory pulp. It can be grown in areas of rainfall ranging from 750-4500mm and is fairly adaptable to different soils. It produces large quantities of seeds, starting in the fourth year, with the result that after the first harvest, a carpet of melina seedlings surrounds each stump and helps suppress

TABLE 4-A-I

Selected Species Worth Considering for Mean Annual Rainfall 650–1000 mm

Mean Annual Temperature (Altitude)

Over 24.5 C (0–500 meters)	*22.0–24.5 C (500–1000 meters)*	*19.5–22.0 C (1000–1500 meters)*
Albizia lebbek	Albizia lebbek	Casuarina equisetifolia
Azadirachta indica	Cassia siamea	Cupressus macrocarpa
Cassia siamea	Casuarina equisetifolia	Eucalyptus botryoides
Casuarina equisetifolia	E. botryoides	E. camaldulensis (proc. septentrionales)
Eucalyptus camaldulensis (proc. septentrionales)	E. camaldulensis (proc. septentrionales)	E. camaldulensia (proc. meridionales)
E. citriodora	E. camaldulensis (proc. meridionales)	E. citriodora
E. tereticornis	E. citriodora	E. gomphocephala
Leucaena leucocephala (Hawaii)	E. gomphocephala	E. maculata
L. leucocephala (Salvador)	E. tereticornis	E. maidenii
Pinus caribaea	Leucaena leucocephala (Hawaii)	E. melliodora
Samanea saman	L. leucocephala (Salvador)	E. paniculata
Sesbania bispinosa	Melaleuca leucadendron	E. tereticornis
	Pinus caribaea	Grevillea robusta
		Pinus elliottii
		P. oocarpa

17.0–19.5 C (1500–2000 meters)	less than 17.0 C (over 2000 meters)
Cupressus macrocarpa	Cupressus macrocarpa
C. torulosa	C. torulosa
E. camaldulensis (proc. meridionales)	E. globulus
E. citriodora	E. st. johnii
E. globulus	E. viminalis
E. gomphocephala	Grevillea robusta
E. maculata	Pinus canariensis
E. maidenii	P. elliottii
E. melliodora	P. halepensis
E. tereticornis	P. oocarpa
Grevillea robusta	P. radiata
Pinus brutia	
P. canariensis	
P. elliottii	
P. halepensis	
P. kesiya	
P. oocarpa	
P. radiata	

Sources: Webb, et al., 1980, pp. 19–20; NAS, 1980, pp. 187–190.

TABLE 4-A-II

Selected Species Worth Considering for Mean Annual Rainfall 1000–1600 mm

Mean Annual Temperature (Altitude)

over 24.5 C (0–500 meters)	*22.0–24.5 C (500–1000 meters)*	*19.5–22.0 C (1000–1500 meters)*
Acacia auriculiformis	Acrocarpus fraxinifolius	Araucaria spp.
Acrocarpus fraxinifolius	Anthocephalus cadamba	Eucalyptus citriodora
Anthocephalus cadamba	Chlorophora excelsa	E. cloeziana
Calliandra Calothyrsus	Calliandra Calothyrsus	Calliandra Calothyrsus
Chlorophora excelsa	Cordia alliodora	E. grandis
Cordia alliodora	Eucalyptus citriodora	E. maidenii
E. citriodora	E. cloeziana	E. microcorys
E. cloeziana	E. grandis	E. paniculata
E. grandis	E. robusta	E. resinifera
E. robusta	E. saligna	E. robusta
E. urophylla	E. urophylla	E. saligna
Gmelina arborea	Gmelina arborea	Pinus elliottii
Melaleuca leucadendron	Melaleuca leucadendron	P. kesiya
Pinus caribaea	Pinus caribaea	P. oocarpa
P. merkusiana	P. merkusiana	
Samanea saman	Samanea saman	
Sesbania grandiflora		
Terminalia spp.		

17.0–19.5 C (1500–2000 meters)	less than 17.0 C (over 2000 meters)
Acacia decurrens	Acacia decurrens
A. mearnsii	A. mearnsii
A. melanoxylon	A. melanoxylon
Araucaria spp.	Araucaria spp.
Cupressus lusitanica	Cupressus lusitanica
Eucalyptus citriodora	C. macrocarpa
E. globulus	Eucalyptus delegatensis
E. grandis	E. globulus
E. maidenii	E. nitens
E. microcorys	E. st. johnii
E. resinifera	E. viminalis
E. saligna	Pinus canariensis
Pinus canariensis	P. elliottii
P. elliottii	P. oocarpa
P. kesiya	P. pseudostrobus
P. oocarpa	P. radiata
P. pseudostrobus	
P. radiata	

Sources: Webb, et al., 1980, pp. 21–22; NAS, 1980, pp. 187–190, 37–38, 62–63.

TABLE 4-A-III

Selected Species Worth Considering for Mean Annual Rainfall over 1600 mm

Mean Annual Temperature (Altitude)

Over 24.5 C (0–500 meters)	*22.0–24.5 C (500–1000 meters)*	*19.5–22 C (1000–1500 meters)*
Albizia falcata	Albizia falcata	Araucaria spp.
Anthocephalus cadamba	Anthocephalus cadamba	Eucalyptus maidenii
Aucoumea klaineana	Cedrela odorata	E. resinifera
Cariniaria pyriformis	Cordia alliodora	E. robusta
Cedrela odorata	Eucalyptus deglupta	E. saligna
Cordia alliodora	E. grandis	Pinus alliottii
Eucalyptus deglupta	E. robusta	P. merkusii
E. grandis	E. saligna	
E. robusta	Gmelina arborea	
Gmelina arborea	Pinus caribaea	
Musanga cecropioides	P. merkusiana	
Nauclear diderichii	Samanea saman	
Octomeles sumatrana		
Pinus caribaea		
P. merkusiana		
Samanea saman		
Terminalia spp.		
Triplochiton scleroxylon		

17.0–19.5 C (1500–2000 meters)	*less than 17.0 C (over 2000 meters)*
Acacia melanoxylon	Acacia melanoxylon
Araucaria spp.	Araucaria spp.
Eucalyptus globulus	Eucalyptus delegatensis
E. grandis	E. globulus
E. maidenii	E. st. johnii
E. resinifera	E. viminalis
E. saligna	Populus deltoides
Pinus elliottii	

Sources: Webb, et al., 1980; NAS, 1980, pp. 187–190.

subsequent weed growth (Hartshorn, 1979, p. 13). It is, however, susceptible to competition from weeds during its first year, and some plantations have been destroyed by browsing livestock. Although relatively free from diseases, it is subject to attacks by some types of insects, in particular, in the the American tropics, by leaf-cutter ants.

APPENDIX 4-B

Species for Arid and Semi-Arid Lands

A number of species have been investigated for use in arid and semi-arid tropical regions. Tables 4-B-I and 4-B-II list particularly promising species, as described somewhat more fully in Webb, 1980.

Many more factors than rainfall and temperature have to be included in any well thought-out species choice. For instance, *Prosopis juliflora* (mesquite) produces a good heavy fuelwood (specific gravity 0.7) in very dry climates, especially in warm climates—some varieties are not frost hardy. It grows in areas with rainfall of 150-750mm, and will yield 75-100 tons per hectare, on a 15-year rotation. (This corresponds to approximately 7-10 cubic meters per year.) It grows in a variety of soils, reproduces readily from suckers (underground sprouts), and competes well with weeds. In fact, its major problem is probably that it is an excessively competitive invader of other plant communities.

Other species can also be too successful. *Acacia saligna*, for example, is an extremely rugged tree that is widely adaptable to barren slopes, wasteland and exceptionally arid conditions in Australia and North Africa (NAS, 1980, pp. 100-101). It is particularly drought-hardy, growing where annual rainfall is as low as 250mm. It tolerates soil salinity, alkalinity, and grows in acid or calcareous sands. It coppices freely. One disadvantage is that its wood is light and resinous. But perhaps a more serious drawback is that, because of its profuse reproductive ability, it has become a serious weed menace, especially by infesting watercourses, and has proved particularly difficult to eradicate.

TABLE 4-B-I

Selected Species Worth Considering for Mean Annual Rainfall 250–400 mm

Mean Annual Temperature (Altitude)

over 24.5 C (0–500 meters)	*19.5–24.5 C (500–1500 meters)*	*less than 19.5 C (over 1500 meters)*
Acacia saligna	Acacia saligna	Acacia saligna
A. senegal	E. brockwayi	Cupressus arizonica
E. brockwayi	Eucalyptus camaldulensis	Prosopis juliflora
Eucalyptus camaldulensis	(proc. septentrionales)	P. tamarugo
(proc. septentrionales)	E. intertexta	Schinus molle
E. intertexta	E. microtheca	
E. microtheca	E. occidentalis	
E. occidentalis	E. salmonophoia	
E. salmonophloia	Parkinsonia aculeata	
Parkinsonia articulata	Prosopis juliflora	
Prosopis juliflora	Schinus molle	
Tamarix articulata	Tamarix articulata	
Zizyphus mauritiana		

Sources: Webb, et al., 1980, p. 18; NAS, 1980, pp. 187–190.

TABLE 4-B-II

Selected Species Worth Considering for Mean Annual Rainfall 400–650 mm

Mean Annual Temperature (Altitude)

over 24.5 C (0–500 meters)	*19.5–24.5 C (500–1500 meters)*	*less than 19.5 C (over 1500 meters)*
Acacia saligna	Acacia saligna	Acacia saligna
A. senegal	Albizia lebbek	Cupressus arizonica
Albizia lebbek	Eucalyptus camaldulensis	Eucalyptus camadulensis
Albizia lebbek	(proc. septenrionales)	(proc. meridionales)
Azadireachta indica	E. camaldulensis	E. gomphocephala
Eucalyptus camaldulensis	(proc. meridionales)	E. tereticornis
(proc. septentrionales)	E. crebra	Pinus halepensis
E. crebra	E. gomphocephala	Prosopis juliflora
E. occidentalis	E. occidentalis	Schinus molle
E. tereticornis	E. tereticornis	
Prosopis juliflora	Prosopis juliflora	
	Schinus molle	

Sources: Webb, et al., 1980, p. 18; NAS, 1980, pp. 187–190.

APPENDIX 4-C

Species for Tropical Highlands and for Temperate Regions

NAS, 1980 treats nine species for the hill lands of the tropics (areas above 1000 meters). Webb et al., 1980 includes information on high altitude species for tropical and sub-tropical climates. A number of these species come originally from temperate regions. Although some species that are observed to do well in temperate climates may also thrive at high altitudes in the tropics, in some cases length of day and light intensity may be the determining factors in success or failure (Webb et al., 1980, p. 1).

More detail on tropical highland species can be found in the references mentioned; the entries in the tables for higher altitudes in Appendices 4-A and 4-B also serve as a list of highland candidates as described somewhat more fully in Webb et al., 1980.

For temperate regions, a number of species have been investigated in the industrial countries for possible fuelwood plantations. A number of genera that have rapid juvenile growth, ease of establishment and regeneration, freedom from major insect and fungal pests, and (for hardwoods) also have the ability to coppice include *Alnus*, *Eucalyptus*, *Platanus*, *Salix* and *Populus* (Fege et al., 1979, p. 359). One of the most interesting cases of tree growing in developing areas, however, comes from a nation in the temperate region: Korea. As mentioned in footnote 9, two *Lespedeza* species, *L.bicolor* and *L.thunbergii*, are leguminous bushes that rapidly cover bare ground and bind soil and also produce a dense, heavy wood (NAS, 1980, pp. 17-20). On a two-year cycle, they reportedly produce about 6 tons in dry weight (or about 9 cubic meters) per year. Other species grown in Korea have been treated in FAO, 1982a; fast growing species that have been successful are two *Populus* (cottonwood) hybrids (*P. x euramericana* and *P.* x *albaglandulosa*) as well as *Paulowni coreana* (FAO, 1982a, p. 33). Besides the *Lespedeza* already mentioned, species especially suited for fuelwood (and erosion control) were *Alnus japonica* (alder), *Robinia pseudoacacia* (locust) and *P. rigida*.

APPENDIX 4-D

Leucaena

An idea of the characteristic that may bear on the problem of selecting a species can be illustrated by reviewing some of the main points made in the specialized reference on leucaena (*L. leucocephala*), NAS, 1977 (see especially pages 1-7 there for a summary).

Leucaena leucocephala is of Central American origin: the name of the Mexican City Oaxaca is derived from an Indian word meaning "the place where leucaena grows." The nature of the plant varies, with some strains (e.g., "Hawaiian") being many-branched shrubs that average only 5 meters in height of maturity, while others (e.g., "El Salvador") are single-trunk trees that grow as high as 20 meters. Yields of leucaena have run from 24-100 cubic meters per hectare per year, the highest yet measured of any tree species (NAS, 1980, pp. 50-51). Leucaena is quite drought tolerant, although it grows best with rainfalls of at least 600-700mm per year (NAS, 1980, pp. 50-51) . It will grow vigorously only at lower altitudes (in Hawaii, at elevations less than 500 meters) and it only yields exceptional amounts of wood per year in fertile, well-drained soil where irrigation or rainfall is adequate. It also is rather intolerant of acid soil.

The seedlings are slow-growing and must be protected from aggressive weeds. The El Salvador type is well-behaved, but the Hawaiian-type leucaena (the bushy variety) becomes a weed itself and has produced dense tangles of underbrush in applications in Guam and in some parts of Tanzania.

Leucaena makes excellent fuelwood and charcoal. It has uncommonly high density and heat value, especially for a fast-growing tree. It readily coppices. It also has many characteris tics useful for nonenergy purposes that increase its total attractiveness as a multipurpose crop. The arboreal variety of leucaena produces stems that can be of useful size for construction lumber, pulpwood, poles, and posts. Up to a point, leucaena forage is highly palatable, digestible, and nutritious for cattle, water buffalo, and goats. Leucaena products must be supplemented with other feeds, however, because the foliage contains a rare amino acid, mimosine, that can be toxic to cattle if consumed over long periods in large amounts. The leucaena can also be used to enrich soils by fixing nitrogen through bacteria on its roots, a common trait of the pea family. In additions, its aggressive root system can break up impervious soil layers. It has also been used in windbreaks and to shade plantation crops such as coffee.

APPENDIX 4-E

Species for Special Problem Situations

Some species grow relatively well in unfavorable environments. The suitability of some popular species such as *Calliandra calothyrsus* and *Leucaena leucocephala* for poor lands has been noted above. Alexander (1982, p. 4) has mentioned the suitability of *Albizia procera* for humid moist lands and *A. lebbeck* for arid marginal lands in the context of Puerto Rico. (Table 4-E-1 shows species that are adaptive to heavy soils, impeded drainage, and alkaline or saline soils.)

Heybroek (1982, p. 15) also mentions some degree of salt tolerance for acacia and eucalyptus, and lists a number of temperate-zone saline-resistance species such as *Populus alba*, (poplar), *Ulmus pumida* (elm), and *Robinia pseudoacacia* (locust). The last two species are also mentioned as candidates for dry soils in cold weather climates. Eucalyptus and *Fraxinus* (ash), *Salix* (willow), and *Liquidambar styraciflua* (sweetgum) are recommended as genera or species resistant to periodic flooding (Heybroek, 1982, p. 15).

TABLE 4-E-I
Selected Species Tolerant of Particular Soil Conditions

Heavy Soils		*Impeded Drainage*
Acacia aurinculiformis	Gmelina arborea	Acacia auriculiformis
A. mearnsii	Leucaena leucocephala (Hawaii)	E. microtheca
A. saligna	L. leucocephala (Salvador)	Prosopis tamarugo
A. senegal	Prosopis juliflora	
Albizia lebbek	Prosopis tamarugo	
Azadirachta indica		
Eucalyptus camaldulensis (proc. septent.)		
E. camaldulensis (proc. meridion.)		
E. globulus		
E. microtheca		
E. occidentalis		

Alkaline Soils	*Saline Soils*
Acacia auriculiformis	Acacia saligna
A. saligna	Casuarina sequisetifolia
Casuarina equisetifolia	Eucalyptus camaldulensis (proc. septent.)
Eucalyptus camaldulensis (proc. septent.)	E. gomphocephala
E. camaldulensis (proc. meridion.)	E. occidentalis
E. gomphocephala	Parkinsonia aculeata
E. microtheca	Prosopis juliflora
E. occidentalis	P. tamarugo
Leucaena leucocephala (Hawaii)	
L. leucocephala (Salvador)	
Parkinsonis aculeata	
Pinus halepensis	
Prosopis juliflora	
P. tamarugo	

Sources: Webb, et al., 1980, pp. 25–26; NAS, 1980, pp. 187–190.

APPENDIX 4-F

Genetic Engineering

by Jerry Teitelbaum

Genetic engineering is the process by which plant and animal cells are manipulated in such a way that the expression of genetic traits can be controlled and selectively chosen to produce an organism with defined characteristics. The ultimate goal of genetic engineering is to produce an improved cultivar or breeding line. Many of the sought-after traits for genetic engineers are complex, polygenically controlled processes such as increased growth rate and yield, improved photosynthetic efficiency, improved biological nitrogen fixation, improved pest resistance, and resistance to environmental stresses (Collins, 1982, p. 246).

The plant breeder must determine what genes are present in a plant genotype (the genetic make-up of a plant) and what traits in the phenotype (the plant's expression of traits) they control. Once the genotype is mapped, the breeder then tries to introduce new genes from exogenous plant sources into the genome (chromosomes). This is done by a variety of technical processes such as using a bacterial vector and enzymatic gene splicing. To determine if the introduced genes will control or enhance the desired trait, the cell must be cultured, mapped, and regenerated into a whole plant. It is important to stress that this process, with our current level of knowledge, is a hit or miss process. We are dealing with extremely complicated processes that must be studied in greater depth and better understood before they become a viable option.

Once the desired trait is proven, the next step is to produce a viable plant that can be transplanted to the field and ultimately produce vegetative progeny or seed that expresses the desired trait. One of the important restrictions on plant regeneration from culture is that the plant that is produced, like the cell culture, loses its reproductive potential over time (Collins, 1982, p. 234).

At present, many of the technologies required to "design" a plant are available. There are refinements of method and carry-overs that must be developed before, however, we can, for example, take genes from plant X that allow it to withstand wide temperature ranges and insert these genes into a cell from plant Y, producing a plant Z that can withstand a range of temperatures. This type of engineering falls in the realm of molecular genetics, an area that is still only imperfectly understood.

Current research is focused on learning more about culturing techniques; the more types of cells that can give rise to normal plants, the greater the potential contribution to plant breeding. Efforts are also being concentrated on understanding the control of the growth of plant cells and tissues (Flavell, 1981, p. 7).

Near-term goals for research, those that will be fully attainable within the next five years, include such areas as clonal plant propagation and wide hybridization. Mid-term goals for research, those that we will be able to develop within 5-10 years, are variant selection and production of haploids (reproductive plants) (NRC, 1982, p. 132).

Over the long term, one hopes to be able to select genes from a gene pool in which all genes are labeled and combine them in such a way that we can select the crop type or types, the average yields of that crop, and give the plant everything it needs to thrive in the environment chosen for that plant. In other words, one should be able to custom design new species of plants.

For the present, some of the more ambitious genetic engineering goals are not yet obtainable. This means that bioenergy planners will have to resort to traditional crops and crop selection. The planner can keep himself informed of general developments in genetic engineering and try to fit these proposed bioenergy crop improvements into future bioenergy projects as new research results warrant, however.

APPENDIX 4-G

Use of Woody Residues in Costa Rica

A useful study has calculated the economically available potential of woody wastes (from trees and shrubs) in Costa Rica (Meta, 1982).

The potential supply in Costa Rica of woody residues from forest clearing is estimated to be quite large. After timber extraction there remain not only the unused portions of timber trees, as mentioned in the text, but more important in this specific analysis, there are existing "trash" trees and shrubs that have not been cleared during timber harvests. Eventually, when the land is converted to agriculture or to pasture land, this clearing will have to be done anyway. But if this clearing were done on an organized basis to gather usable biofuel, some 900 thousand tons of woody residues would be available, not only for home cooking and similar uses, but also for industrial burning of wood or for gasification (Meta, 1982, pp. 2-5, 2-14).

There are also the woody residues from sawmill operations mentioned in the text. About 200 sawmills in 1980 processed 894,200 cubic meters (solid basis) of roundwood, of which approximately 45 percent were discarded as residue (Meta, 1982, pp. 2-5, 2-8ff). Of this residue about 75 percent are slabs that can be chipped or made into charcoal; 25 percent are sawdust and shavings that can be either used directly, pelletized, or converted to charcoal and pyrolytic oil. It was calculated that biofuels from this source could sell for $3 a ton or less. Finally, informal evidence indicated that a very large source of present fuelwood supplies came from residues from trees planted for other purposes (such as shade trees for coffee, cocoa, etc.). The current domestic market supplied from this source was estimated as three times larger than the size of the market needed as an outlet for a proposed program retrofitting industrial firms to biomass energy over the next five years (Meta, 1982, pp. 2-7ff). But the potentialities of the "shade tree" source evidently require considerably more investigation since quantitative market data are lacking.

5

Environmental Impacts of Bioenergy Crops

Introduction: The Environment and Site Suitability

The environmental consequences of a bioenergy project may be quite complex—and sometimes indirect.[1] One of the greatest environmental impacts of present patterns of *irrational* use of traditional fuels is to make some contribution—sometimes small, sometimes large—to the crisis of deforestation (Allen and Barnes, 1981) or to the diversion of crop residues from soil improvement to fuel uses. But a planned program of biomass cultivation should act to relieve, rather than intensify, current pressures on natural forest resources and the environment.

The *direct* environmental effects of a bioenergy crop depend on what the crop is and where it is planted. Effects should be small or nonexistent if sugarcane presently grown for sugar is instead turned to alcohol production, for instance. And new energy uses for residues like sugar bagasse for fuel should lead only to environmental impacts that can be readily controlled. But the planting of new types of field crops on existing arable lands or on new lands, or the establishment of forest crops on the sites of former natural forests, wastelands, or on pasture lands, will inevitably cause changes in the local environment. These may be minor changes, easily accommodated within the existing ecosystem, or they may be major changes drastically altering the site or its surroundings. Some of the complex changes in local communities of plants and animals that may occur could be considered desirable and others undesirable—depending on the specified economic, social, and "conservation" values accepted (consciously or unconsciously) by both planners and local inhabitants.[2] The main point to emphasize is that no kind of bioenergy production, even the gathering of fallen wood from natural forests, is devoid of environmental

1. This chapter is largely drawn from Allen and Davis, 1983; this work is a reference for much of the otherwise unattributed material below.

2. These values or criteria may be much more subtle than sometimes realized. For example, see Ramsay, 1972, Chapter 6 and Ramsay, 1979, Chapter 12.

implications; at the same time, environmental impacts can produce good as well as bad results.

We discuss here the direct environmental implications of some typical kinds of bioenergy crop production. These environmental effects occur in three places: on the crop site itself, in areas adjacent to the site, and especially—given the importance for regional watersheds of effects like erosion —in downstream locations. The nature and degree of the impacts vary over the bioenergy production cycle of planting, tending, and harvesting. Impacts also necessarily vary with the type of crop produced.

Abatement strategies are also important to the bioenergy planner: environmental impacts can be substantially mitigated if the bioenergy cropping process is properly designed. Finally, a few generalizations can be made about the suitability of different types of sites—how to select sites to minimize negative environmental impacts and to encourage positive ones.

The Site As A System: Characteristics and Environmental Vulnerability

The environment of a bioenergy site is characterized by rainfall, soil, and the existing biomass community. All of these elements participate in two fairly complex and interacting cycles, a hydrological cycle involving water movement, and a biogeochemical cycle involving soil formation, plant growth, and everything else.[3] Biomass growth—that is, the formation of new plant tissue out of the raw materials, carbon dioxide, water, and miscellaneous soil nutrients—depends inherently on the detailed workings of these two cycles. Any new bioenergy project can be expected to modify the cycles in one or more places and therefore to present an environmental problem—or opportunity. When disturbances occur, they express themselves both through changes in the existing biomass cover (cutting down an existing forest, for example) and in direct changes to the soils. Changes in the biomass cover, in turn, indirectly cause additional soil changes.[4] Finally, the cycle of rainfall and other water movement will be affected by changes in plant cover and, especially, by modifications to soil structure.

The soil is therefore the linchpin in the biomass-soil-water system, and the properties of soil have been subjected to a great deal of

3. The biogeochemical cycle encompasses the weathering and breakdown of rock, litterfall from vegetation, soil formation, release of nutrients in soils, nutrient uptake, and plant growth. The hydrological cycle involves rainfall, interception, throughfall, stem flow, infiltration, run-off, soil moisture storage, ground water recharge, water uptake by plants, evapotranspiration, and evaporation.

4. Soil is composed of both a mineral fraction—weathered rock materials—and an organic fraction—decomposing or decomposed plant and animal material; therefore changes in the biomass cover due to a new bioenergy project show up sooner or later in soil modifications.

scientific analysis. The effect of different quantitative and qualitative changes in the key properties of soil—nutrients and structure—on the growth of crops have also been widely studied.[5] These studies stress the following points: first, bioenergy cropping generally will be constrained by the limited availability of one or more essential nutrient elements—the exact elements involved and quantitative needs will vary with the species (Jordan, 1977, 1978; Nye, 1961, pp. 35-42). The reason for the nutrient deficiency in the soil may be of practical importance as well, because countermeasures can be taken for some problems and not for others.[6]

The structure and texture (e.g., coarseness, cohesiveness) of the soil are also crucial: they determine the amount of rooting space and the soil's capacity to hold water. Soil structure and texture also control whether water and nutrients are transferred efficiently to the roots of the plants and determine how much water is retained and how much runs off and is lost off-site.[7] Some soils have relatively high infiltration capacities and are therefore able to absorb water more quickly and transfer water better. On the other hand, heavier soils with large fractions of clay are able to retain more total water.[8] As we shall see, water transport and carrying capacity will often be changed significantly by new bioenergy production schemes.

Impacts From Biocropping

Clearing a site and planting, tending, and harvesting a bioenergy crop produces impacts that may be lasting—or that may be repaired by natural processes as time goes on. These impacts vary, of course, depending on what crop is planted, and their pattern extends outwards from the site into surrounding and downstream regions.

---

5. The main soil chemical properties affecting yield are pH, organic content, and available nutrients, especially phosphorus, potassium, calcium and magnesium. Soil acidity or pH (the negative log of the concentration of H ions) reflects the degree of weathering of the mineral fraction of the soil and the nature of the vegetation growing on it and may also indicate the amount of leaching of bases from the soil; these bases and exchangeable cations (positively charged particles) include atoms or ions of potassium, calcium, magnesium and sodium. Certain anions (negatively charged particles), e.g., phosphorus and sulfur, are also essential to plant growth. The anions and cations in soils may come from atmospheric sources or decomposed organic matter or may be released by weathering from minerals.

6. (1) Soils may be inherently deficient in a particular element, or (2) previously adequate stores of that element may have been exhausted by weathering or by some particular land-use practice, or (3) available reserves of the element may not be getting transported efficiently to the plant roots. If a soil is *inherently* deficient in a certain nutrient, that nutrient may have to be added to a fertilizer if exotic species are to be cultivated. But if the soil has been *exhausted*, nutrients may reaccumulate from the air, the rain, and by accumulation of soil litter. If on the other hand nutrients are not being transferred to the plants fast enough, one can sometimes add more water in order to accelerate mineral and organic breakdown and stimulate the transport of nutrients.

7. The amount of water reaching the soil depends on the amount that passes through the plant canopy as throughfall or stem flow, rather than being intercepted, plus amounts from nearby surfaces from rainfall or irrigation.

8. Soils with "good" structure or coarse texture are typically those with large sand fractions or with many aggregates of clay particles held together by free iron or aluminum ions or by certain organic structures (metacolloids). Soils with less desirable structure have a higher ratio of capillary and osmotic pore space to gravity pore space and are usually characterized by having larger clay fractions. (Allen and Davis, 1983, p.9).

Site Clearing and Preparation and Its Consequences

One of the biggest impacts of a bioenergy project is the clearing of the site: removal of vegetation, plowing, and application of pesticides or herbicides.

The main effect of clearing the existing vegetation is that the natural cycle of return of nutrients to the soils is sharply interrupted and the soil is exposed to direct rainfall and sunlight. In addition, any heavy equipment used will cause increased soil compaction, while ploughing and stump removal can affect soil characteristics radically.[9]

After clearing, more rainfall reaches the soil. Furthermore, the infiltration and holding capacity of the soil may have been reduced by direct disturbance or the removal of the normal protective layer of litter. Because of both higher raindrop impact and reduced infiltration, a larger fraction of the rainfall will not penetrate the soil, but instead will run downhill as overland flow. Surface runoff commonly carries soil particles with it and may cause sheet, rill, or gully erosion.[10]

When erosion of upper soil layers is severe, it can expose a subsoil, that, more often than not, has a higher clay content and less organic matter than the surface soil. Thus the soil now has a lower infiltration capacity, increasing the proportion of future rainfall that runs off—and so a destructive feedback cycle may be created. The possibility of wind erosion is also often increased.

The removal of the existing vegetation also disturbs the local community of plants and animals. Especially affected are the small animals, insects, and microscopic creatures that play an essential role both in making nutrients available to plants and in altering physical properties of the soil: moles, earthworms, termites, and other burrowers loosen the soil and make otherwise inaccessible nutrients available to plants (Allen and Davis, 1983, p. 11). Bacteria are especially important; in particular, some species of bacteria decompose organic matter and "mineralize" the nitrogen in the organic matter into a form that is available for use by plants (Black, 1968, pp. 405-557; Russell, 1973, pp. 327-787).[11] Obviously, the disturbance of such a complex ecosystem will change the nature of the local soil and therefore the environment.

---

9. Stump removal creates large holes that can increase infiltration of rain water, add to soil compaction, and expose less permeable soil layers.

10. "Sheet, rill, or gully" refers to the uniform, streamlet, or small stream appearance of the erosive flow, respectively.

In some cases the increased rainfall saturates layers of soil, creating a slurry of water and soil particles that may appear suddenly as a slump or landslide. (Allen and Davis, 1983, p.10).

11. Other bacteria—especially those associated with the root systems of members of the pea family—"sift" nitrogen from the air. Still other bacteria, together with certain fungi, improve soil structure by forming aggregates of clay, silt, and sand particles (Allen and Davis, 1983, p.11).

The net effects of land clearing may often be negative but sometimes are not. Clearing may result in the temporary increase of some nutrients, for instance, an increase in nitrogen from the decay of litter.[12] This will help the regrowth and recovery of vegetation on the site. But the typical result is a deterioration of important soil-building processes. We will look at the soil modification problem later, after first examining the effects of bioenergy management actions that are taken after the clearing of the original vegetation.

Bioenergy Management Actions After Clearing

Clearing is apt to be the most ecologically traumatic event in the life of a new bioenergy site. But subsequent actions —planting; intermediate treatments such as tilling, weeding, and application of fertilizer; and harvesting—will determine how the soil and the site in general will subsequently recover.

Planting. Crops can be transplanted or seeded. Both procedures have disadvantages as far as soil disturbance is concerned. Digging holes for transplanting has some impact on the soil. A more serious impact is the exposed soil left between the seedlings themselves. If seedlings are spaced fairly wide apart, however, the initial clearing of vegetation need not be complete.

In contrast, direct seeding may well require complete vegetation removal, so that for a time soil will be exposed and subject to erosion. In addition to erosion problems, the rate of evapotranspiration from the soil will in general be changed; however, the degree to which changes in evapotranspiration affect the rest of the hydrologic cycle is controversial (Allen and Davis, 1983, p.19).

Intermediate Treatments. Weeding and thinning out of seedlings are the main mechanical treatments that can affect the amount of litter added to the soil, the amount of throughfall of rain, and exposure of the soil to sunlight. Bioenergy planners may want to consider the soil-protection advantages in retaining litter from these operations.[13]

Chemicals also cause problems. Application of fertilizers and pesticides can lead to nutrient "overload" in local watercourses and to the dissemination of toxins to offsite locations.

Harvesting. Impacts from soil disturbance are proportional to frequency: the more frequent the disturbance, the less time for soil recovery. Seasonality is also important; if the soil is left exposed during

12. Increases in soil temperature, moisture and litter (assuming the litter is not removed) on the soil surface after clearing can lead to a sudden increase in populations of "heterotrophic" microbes that will mineralize nitrogen, making it available in soluble form.

13. Note, however, that weeds left as residue will tend to reseed and that decomposition of the litter often is accompanied by losses rather than gains of nutrients in the short term as well as in the long term.

seasons of high rainfall or extremely dry, wintry conditions, risks of erosion will increase. The method of harvesting is also important. As with clearing, the use of heavy machinery for harvesting will tend to compact soils and increase runoff. The amount of litter left on the site will affect infiltration and runoff and of course eventually the nutrient balance. Incidentally, there are possible tradeoffs between removals of bioenergy for fuel and maintenance of nutrients on the site—the parts of the plant that have high cellulose or other carbohydrate content are not usually those parts with the highest nutrient content.

How Soils Change After Clearing and Cropping

Clearing natural vegetation from a site tends to change the structure and nutrient levels in such a way as to decrease soil productivity. This not very surprising result can be expressed in scientific terms: such important measures of soil fertility as percentages of organic carbon and total nitrogen have been observed to decrease after the conversion of tropical forests to either plantations, pastures, or field crops (Allen, 1984, pp.24-26). The lightness of a soil, as measured by "inverse bulk density," also decreases significantly, reducing its structural qualities.[14]

These correlations are not without their complexities. "Plant available phosphorus," for example, tends eventually to return to original levels after deforestation. It experiences a sharp *increase* very shortly after deforestation, as a result of increased decomposition or burning of organic matter. This source of phosphorus is, however, soon depleted. The phosphorus level therefore merely subsides over time to its original (preclearing) low levels.[15]

One measure of the efficiency of transfer of elements within the soil, the "cation exchange capacity," is decreased considerably after clearing and replanting, especially for certain types of soil.[16] In fact, the decreases in organic carbon and nitrogen after clearing are much greater for "old-parent material soils," formed on highly weathered rock or colluvial deposits, than for "young-parent material soils,"—formed on recent volcanic deposits or newly exposed plutonic rocks (Allen, 1984, p.32; Allen and Davis, 1983, p.17).

The moral for bioenergy planners is that one must expect some permanent losses in the natural balance of certain types of desirable

14. The precise results are discussed fully by Allen and Cady, who cited a range of cases comparing different types of tropical and temperate soils. They give statistical analyses showing significant decreases in magnesium, increases in exchangeable calcium and in pH, with in general no effect on exchangeable potassium (Allen, 1984, p.29). Some of these results differ according to soil type (see below in text).

15. Over time, the available phosphorus is generally taken up by oxides of iron and aluminum that capture it into forms not readily available to plants.

16. Cation exchange capacity (CEC) is the ability of the soil to hold nutrients for plants; the higher the CEC of a soil, the more effectively it can hold fertilizer.

soil components after conversion of the original forest to tree plantations or to field crop uses. Some properties such as potassium levels, however, are not extremely affected and others can actually be rebuilt, soil "lightness" (inverse bulk density) for example. Finally, the geological origin of the soils (weathered versus unweathered material) may be as important or even more important than climatic conditions or crop management in predicting soil recovery after clearing (Allen, 1984, p.37).

The Effect of Different Crops

The problems mentioned above in the section on clearing can occur in all kinds of bioenergy projects. The problems in the sections treating actions after clearing, while modeled specifically on practices in plantation tree farming, are also common to many field crops. Certain special environmental problems, however, are more significant in connection with specific kinds of bioenergy crops.

Plantation Forests. We have already examined the general soil structure, soil nutrients, and water-related problems of typical plantation management systems. But the use of herbicides for weed control also presents potential hazards. These herbicides, once their local function is completed, may persist in soils and waterways and cause significant ecological damage downstream to animals as well as plants—the half-life for the most commonly used herbicides averages between 15 and 130 days.

Marginal Forests. Removing marginal forests has much the same basic effect as removing any kind of original biomass cover, as sketched above. But forests that are marginal tend to occupy sites unfit for commercial timber or crop production: sites with steep slopes or poorly drained or otherwise defective soils. Cultivating biocrops on such sites may have unusually serious impacts on soils and therefore on local hydrology.[17] For example, the removal of vegetation, particularly trees providing shade along stream banks, can raise average stream temperatures several degrees Celsius, thereby lowering dissolved oxygen levels (Allen and Davis, 1983, p.23).

Forest Residues. When one uses logging residues as a bioenergy source, nutrient balances are directly threatened. Forest residues (litter) typically have nutrient concentrations 10 times those of the bole (trunk) wood (Duvigneaud and Denaeyer-de Smet, 1970, pp. 125, 199). The

17. Removal of tree cover increases the evaporation and decreases transpiration from the site and may increase overland flow as much as 15-25 percent. (See for example Kockenderfer and Aubertin, 1975). Deforestation has been estimated to produce a 5 to 25 fold increase in sediment (Dunne and Leopold, 1978) and a 2 to 10 fold increase in nitrogen and phosphorus yield (Likens and Bormann, 1978). The actual increases are a function of slope, soil texture, water holding capacity, and the rate at which the vegetation cover regenerates (Allen and Davis, 1983, p.22).

importance of corrective planning—through selective harvesting or the addition of nutrients—cannot be overemphasized.

Removing the residues from sites harvested for timber may also slightly increase water runoff by reducing the surface area for evaporation and by compacting the soil, thus decreasing the infiltration of rain water. Where sites already possess a topsoil rich in nutrients, removal of residues may lead to an even higher nutrient level in streams as organic matter and nutrients absorbed on soil particles are delivered to the streams (Allen and Davis, 1983, p.22).

Agricultural Residues. The situation for agricultural residues is similar to that for wood residues. There are certain differences, however. The problem of wind erosion may be much greater with field crops, and residues from field crops can generally play a more critical role in prevention of erosion than those from forest crops. Oddly enough, there are also some cases in which removing residues may benefit the environment. Residues from corn, sugarcane, and wheat may release toxic substances during decay and hence lower site productivity (Elliot et al., 1978).

Agricultural Energy Plantations. The fertilizer problem is especially critical for agricultural energy plantations such as those growing sugarcane for fuel alcohol production. Field crops require relatively large amounts of nitrogen and phosphorus, and runoff of these elements is of special concern because they often regulate the productivity of algae in aquatic systems.[18]

The insecticide problem is especially severe for energy field crops. The bioenergy planner may have to take special account of application methods: local waterways are especially vulnerable when application is made either from low flying aircraft or immediately preceding rainstorms.[19]

Freshwater Biomass. A bioenergy system for growing aquatic species like water hyacinth or various types of algae, despite its superficial attractiveness on other grounds, could have troublesome impacts on local hydrology. The balance of nutrients—in a minimum input scenario—will be changed by the harvesting of bioenergy aquatic crops even when the species occur naturally in local communities. If exotic crops are selected or aquatic environments are artificially created, major imbalances could appear in the ecosystem. In practice, most

18. The mechanics of pollution may be different for the two elements. Nitrogen occurs principally as water-soluble nitrate. Much of this nitrate will be lost to local streams if the fertilizer is added before crop root growth is advanced enough to take the nitrates up in quantity. Phosphorus, on the other hand, occurs bound in the organic fraction of the soil, moving into streams attached to eroded soil particles. Phosphorus loading is therefore more easily controlled by good agricultural soil management practices (Allen and Davis, 1983, pp.25-26).

19. The mobility of the insecticide depends on its chemical properties and also on ambient moisture and temperature, acidity, organic matter, and site factors related to top soil erosion (Allen and Davis, 1983, p.26).

aquacultural schemes will require the use of additional nutrients; but added nutrients, as we have seen above in other contexts, can be too much of a good thing. The construction of ponds or the modification of existing waterways for cultivation of aquatics can obviously affect local hydrology also.

Offsite Problems

Effluents of one kind or another from bioenergy sites cause offsite problems, as we have seen.[20] The most common pollutants —suspended sediment, crop residues, excess nutrients, herbicides and pesticides—can inflict a number of different effects on offsite locations.[21] Suspended sediment can cause significant effects on local watercourses and therefore on downstream agriculture. The effect of increased erosion and transport of sediments on downstream reservoirs is also a crucial and complex question. If bioenergy cropping substantially increases sediment transport rates, it could cause extensive damage to downstream dam projects. On the other hand, it is possible to use bioenergy planning to decrease the siltation of reservoirs and the consequent consequent loss of irrigation water. In addition, if it diminishes sedimentation of reservoirs, bioenergy cropping could serve to improve flood control potential and hydropower capacity.[22]

Discharges of residues from clearing or cropping compete for oxygen in local watercourses, while nutrient overloads from leached fertilizer or eroded soils tend to favor some plant species—such as blue-green algae—at the expense of others.[23] Herbicides and pesticides, as we noted, retain their toxicity offsite and can be poisonous to plants,

20. In addition to conventional pollutants, changes in vegetation at the site can cause changes in the local microclimate and therefore may have effects on rainfall and winds on the site elsewhere. However, it is difficult to assess the significance of these effects: in most practical cases, they should probably be assumed to be small.

21. Sediments and residues can cause a wide variety of impacts to the environment, as has been publicized extensively over the past two decades. (See for example Ramsay, 1972, especially Chapters 10 and 11.) As a reminder, note that suspended sediment in streams lowers water quality for drinking, reduces net photosynthesis by algae and aquatic plants, impairs feeding by filter-feeding and bottom dwelling stream organisms, and may clog the gills and bury spawning beds of fish. Toxic metals and some nutrients such as phosphorus enter streams attached to fine sediment; reservoirs downstream trap such sediment, and as the attached chemicals are subsequently released they will affect algae productivity and related biological processes and may change the species of algae present. Excess algae populations can also lead to oxygen depletion that can locally reduce fish populations. Nitrate-nitrogen concentrations greater than 7 to 8 parts per million can lead to nitrite poisoning in both humans and livestock (Waldbott, 1978).

22. A scheme has been suggested, for example, for the Roseires Dam in the southeast part of the Sudan, to combine bioenergy with watershed retention (Sudan, 1980, p.8). Siltation of the reservoir is proceeding five times as fast as anticipated, and it was hypothesized that the annual cropping of the banks of the reservoir was a significant contributor to this problem. Siltation islands in the reservoir have led to temporary losses of 65 megawatts of power, and the threat of permanent loss of capacity. A water-compatible tree species such as *Acacia nilotica* could assist soil retention and could be harvested to supply fuel and charcoal. A rough estimate of the cost of the charcoal from such a project, not taking into account any credit for the watershed management, would be about $10 per ton or 50 cents a gigajoule.

23. We have considered blue-green algae in the text above (see Chapter 4) as a possible exotic energy crop. In most conventional contexts, however, they combine ecological aggressiveness with economic unattractiveness.

animals, and people downstream. Furthermore, the most commonly used pesticides are extremely stable in natural environments, persisting for weeks, months, or even years.

The bioenergy planner cannot be held solely responsible for initiating and overseeing complex analyses of these very involved offsite pollution problems. But both the dangers of pollution —if project management is ill-planned—and the possibility of environmental "repair"—when environmental consequences are intelligently anticipated—must be kept in mind in assessing biomass energy schemes.

The Human Dimension

The growing of new bioenergy crops will affect not only the physical environment, but also the people involved in the bioenergy labor force. The possible impacts of toxic pesticides and herbicides on humans has been noted. Other health and safety risks involved may not be negligible. Indeed, one study for the United States shows that forest biomass industries have several times more occupational injuries and illnesses than coal mining and oil drilling (Pimentel et al., 1982, p. 24). Conditions in developing countries, however, are different; in particular, less use of mechanical equipment should lower safety risks. Nevertheless, measures to anticipate and mitigate these safety risks cannot be ignored in planning evaluations.

Abatement Measures For Impacts

Even when bioenergy projects cannot be designed to produce net environmental benefits, adverse environmental impacts can be mitigated or abated. An assortment of abatement measures can be taken against stream pollution and other environmental impacts from the bioenergy site.

Perhaps the most effective and most "natural" method of abating environmental impact is to maintain continuous ground cover to whatever extent possible. Such maintenance will reduce both leaching of nutrients and soil erosion. After a natural cover of trees has been cut, for example, a new cover of herbs or grass will help to rebuild the soil. Even the retention of organic litter or other wastes on the soil will retard soil losses.

This kind of reasoning, in fact, suggests that agroforestry can have environmental advantages, in addition to its attractive economic features.[24] If field crops are maintained in agroforestry projects after the

24. Agroforestry consists of the growing of annual field crops such as beans or corn in among the saplings in a tree plantation.

forest crops are harvested, considerable soil retention will take place. Even if the field crops have to be harvested before or shortly after the tree crop harvest, the roots or the litter of the field crops, if retained in or on the soil, can serve a useful soil stabilization function.

There are other management measures that can be taken to reduce negative impacts. In some tree planting schemes, tree stumps may be retained for coppicing purposes; but even when coppicing is not a consideration, erosion will be diminished if the tree stumps are left in the ground.

Another way of minimizing sediment discharged into streams is to maintain unharvested buffer strips, say, 20-40 meters wide, along local waterways. These strips will greatly reduce the entry of residues and associated chemicals into local waterways.

For runoff channels formed by natural depressions or hillside gullies, grasses such as Bermuda grass or stem grass (*Cynodon* spp.) can be planted to stabilize them (Brewbaker, 1980, p. 42). A great deal can be done with road design. Roads into harvesting sites can be constructed to follow hillslope contours as much as possible. Drainage ditches lined with vegetation or rough stones placed along the road can help to lower peak water discharges during storms and to minimize the export of sediments from nearby surfaces and from the roadbanks.

Another useful abatement device is to dig sediment basins near the base of upland slopes. Such basins are normally designed to store a 10-year frequency run-off,[25] and are equipped with an emergency spillway at either end to handle larger storms (Allen and Davis, 1983, pp.31-32).[26]

Planning Procedure: Environmental Siting Criteria

Just as the biological site selection process for a successful bioenergy crop is not an exact science, so too it is impossible to foresee all local and downstream environmental consequences before the fact. One major factor that *can* be predicted, at least in principle, is erosion or water runoff. It is observed that water runoff increases significantly after the cutting of the natural cover, but the increase varies widely between different locations and different crops.[27] Some studies seem to show that, if the initial damage has not been too great, the soil tends to return to normal water management characteristics after a reasonable number of years.[28]

---

25. The maximum flooding that would be expected to occur during any 10-year interval.

26. These basins can be very inexpensive, consisting of nothing more than earthen embankments and spill pipes or weirs (Beasley, 1971).

27. Some of the results for water losses for various locations and soils can be found in Dunne and Leopold, 1978.

28. It has been observed that initial increases in runoff after clearing subsequently "decay" exponentially with a "half-life" of up to seven years (Dunne and Leopold, 1978).

The behavior of a given soil can be analyzed as depending on four different kinds of factors: (1) soil susceptibility to erosion; (2) rainfall intensity and duration projections; (3) proposed cropping, and; (4) management strategy. With a knowledge of these factors one could *in principle* predict the relative erosive environmental damage to a site. The Universal Soil Loss Equation (USLE) is an empirical equation that takes into account a number of these factors.[29] The "universal" equation, however, is considerably less than universal, and the calculations cannot be reliably used to predict *absolute* amounts of erosion. Nevertheless, such equations, with modifications for local conditions, could be used by bioenergy planners to indicate the relative severity of erosion.

Finally, a useful generalization on siting criteria is that the largest increases in water runoff can be expected from harvesting woodlands on steep slopes, or on loose granular soils, or in areas of high rainfall *and* high rainfall intensity (high rainfall "impulse," or volume per unit time). It is therefore appropriate for bioenergy planners to closely examine the siting of projects when one or the other of these factors is apt to be present.

Conclusion

Siting for bioenergy cultivation will inevitably involve environmental impacts. With careful planning, however, impacts can be good as well as bad. Site clearing will usually produce the largest impacts on the site because of impacts on soils, but the processes of planting, intermediate treatments, and harvesting should also be examined to design environmental protection measures. The disposition of residues after harvesting plays an important role, especially for field crops.

The susceptibility of soils to permanent damage at a significant level depends strongly on the geological origin of the soils. Erosion can be minimized by proper site selection— taking slope angle and length and rainfall intensity into account —and by conservation measures like buffer strips. These factors must be taken into account both in designing biomass cultivation procedures and in evaluating sites with environmental criteria in mind.

---

29. Soil loss equations take into account, particularly, the slope angle and length associated with the soil, the amount of rainfall intercepted by the crop, the energy and the timing associated with the rainfall (i.e. the impulse generated by the rainfall), and empirical estimates of the susceptibility of the soil to sheet and rill erosion. (See Dunne and Leopold, 1978).

6

Bioenergy Conversion Technologies

Introduction: Biofuels and Their Production

One can safely assume that future research will produce revolutionary new technologies for bioenergy. These technologies should lower costs and raise efficiencies for turning basic biomass feedstocks from woody and herbaceous plants directly into energy—or for converting them first into more efficient or convenient modern biofuels that can be later burned to produce useful energy. But there is no need to wait for future wonders: a number of efficient technologies are already available. These present technologies can be divided into two economic "classes." One class is already "commercial"—meaning that costs and reliabilities are well known. But such commercial technologies produce fuels that in most places in the world are still more expensive than competing fossil fuels. The other class of newer bioenergy technologies has not yet been demonstrated to be commercially practicable. Such a lack of commercial credentials must be considered a drawback in considering them for present-day bioenergy projects. On the other hand, with further experience and improvements, some of these newer technologies might turn out to produce lower-priced energy than their fossil equivalents.

In this chapter, we try to match the kinds of energy end use products needed in a country's demand picture with the bioenergy fuels presently used and with newer bioenergy fuels that might turn out to be better substitutes for petroleum and other fossil fuels.[1] Then we briefly review the general outlook for the most common technologies, or ways of producing these products, and how they relate to feedstocks of different kinds. Finally, we try to show how to integrate informally this outlook for products and technologies with the conclusions of the previous chapters on land use and bioenergy crops. One can then make a preliminary feasibility assessment for a particular region or country. The assessment is "preliminary" because feasibility will be

1. Newer biofuels can of course substitute for traditional biofuels, e.g., biogas for the direct use of animal wastes.

based only on three key factors—land use, crop possibilities, and energy conversion technologies. Other factors—social, environmental, organizational—must still be considered, as later chapters will show.

Energy End Uses: The Role of Biofuels

Energy end uses can be conveniently divided into several categories. Here we examine the broad categories of motor fuels, household uses, and process heat (industrial heating needs) in a wide variety of rural biomass energy options.

Motor Fuels

The ideal kind of bioenergy fuels to replace gasoline and diesel fuel—in running internal combustion engines for transport, industry, or electricity generation—would be more or less exact replicas of petroleum products, made synthetically from wood or other biomass. A certain amount of experience has been accumulated in making short-chain hydrocarbons (the kind mostly used in motor fuels) from coal, and similar processes could use biomass instead of coal as a feedstock. But for reasons of cost and scale, such synthetics are probably not a good bet at present for most developing countries.

Internal combustion engines can also be fueled by either wood alcohol (methanol) or grain alcohol (ethanol), which can be made from biomass by various means. Unfortunately, alcohols are not perfect substitute fuels for gasoline engines (Dunkerley et al., 1981, pp. 197-198). True, blends of 10 to 20 percent alcohol with gasoline will operate in conventional engines without change. These blends (gasohol) have disadvantages, such as problems with the tendency of alcohol to absorb water,[2] but such drawbacks are far from crippling. On the other hand, to significantly reduce national petroleum demand, not just a middling fraction of it, pure alcohols are a much more promising biofuel. They too have disadvantages: as 100 percent substitutes for gasoline, the alcohols require some modification of carburetors and intake systems because the volatility characteristics of alcohols are different from those of gasoline. The alcohols also tend to corrode certain plastics and metals commonly used in automotive engines, and usually lubrication methods must be changed in order to counteract tendencies for alcohols to cause excessive engine wear.

Another disadvantage of the alcohols is that they cannot be used as is for diesel fuels. A mixture of half diesel and half alcohol is a possible

2. The water absorbed tends to make different gasoline fractions separate out, making the fuel inhomogeneous and ignition erratic.

compromise, or additives can be used with pure alcohol in diesel engines to improve ignition characteristics. Or the diesel engine can be modified in the direction of the usual (Otto cycle) gasoline engine configuration so that almost pure alcohol can be used.[3]

These problems with alcohol as a replacement for diesel suggest considering vegetable oil fuels. As we noted in chapter 4, some of the earliest demonstrations of diesel engines used peanut oil; and soybeans, sunflowers, palm oils, coconuts, castor beans, and sesame seeds could supply a valuable diesel motor fuel substitute. This does not mean that problems do not arise. In some tests of vegetable oils, injection nozzles have at times clogged and residues have accumulated inside the cylinder (Hall et al., 1982, p. 55). It is reported, however, that such problems can be overcome with minor engine modifications, or pure oils (esters) can be derived from the crude vegetable oils in order to produce viscosity and distillation characteristics closer to those of diesel oil.[4]

Gaseous fuels from biomass can also be used in internal combustion engines (Hall et al., 1982, pp. 74-75). The gas must be cooled and relatively free from tar to prevent engine fouling[5,6], and varying degrees of engine modification are generally needed.[7] While gasifiers were in fact used during World War II for vehicle engines, there is the obvious problem of vehicle space requirements and additional weight both with the gasifier itself and with the feedstock supply. Although most European experience with gasifiers has been with the "producer gas" derived by heating biomass in an air-poor environment, biogas from anerobic (oxygen-deficient) fermentation of methane can also be used in internal combustion engines.[8]

Household Cooking, Space Heating, and Lighting

Household energy in the Third World is derived from as many different fuels as in industrial countries, and indeed even among the

3. In fact, if a separate ethanol intake is introduced into the air intake, a diesel engine reportedly could use as much as 85 percent alcohol of 150 proof or better (BOSTID, 1981, p. 90).

4. Costs of this "trans-sterification" process, however, have been reported to be high—$200/tonne (Trindade, 1984). Char-oil (charcoal-petroleum) slurries could also conceivably be used in diesel engines (MITRE, 1981, p. A-30, 31).

5. The main trouble with recent cases of producer gas use has been noted to be engine deposits, and the cure might be preheating the air or more cooling of the gas before it enters the engine (Kristoferson and Kjellstroem, 1981, pp. 1-8).

6. Charcoal is a better gasifier feedstock for this purpose than wood: volatilized tars from wood may have to be cleaned from the gas for internal combustion applications, thus causing a loss of heat and efficiency (Watt, 1982, p. 182).

7. The simplest method here may be to use diesel oil together with a biomass gasifier to fuel a diesel engine, where under normal conditions, the diesel oil is only required for engine starting and warm-up (Hall et al., 1982, pp. 74-75).

For example, Indian trials are being carried out on a generator system that will operate either on diesel or on a 70/30 biogas/diesel mixture (Chopra, 1984).

8. The Stirling (steam) engine could possibly replace internal combustion engines in some uses but is not yet commercially established (Kjellstroem, 1981, p. 10).

lower income classes urban (and sometimes rural) fuels often include gas (piped or bottled), coal, and electricity (Cecelski et al., 1979, Table II-2, p. 16). On the other hand, the most common fuels used the world over in isolated rural regions are wood or wastes and, to a somewhat lesser extent, charcoal and kerosene. Indeed, in many countries, the main "bioenergy problem" may be to keep a supply of biomass (wood) flowing to satisfy the cooking needs of the population. Therefore, a multitude of bioenergy projects could and do involve the conceptually simple—but often practically difficult—task of assuring an adequate bioenergy fuel supply in the form of traditional fuels from wood or wastes. Another large class of projects works to help conserve traditional biofuels by improving the efficiency of combustion of household stoves.

What about substituting biofuels for household fuels presently used? First, there has been a great deal of interest in substituting *newer* forms of bioenergy for older types such as crude stoves or fireplaces using wood, charcoal, or crop or animal wastes. A very straightforward, traditional substitution is to use more charcoal and less wood. Usually this involves no great technological problem, only problems in efficiency and cost—when the source of wood is nearby it may not pay to incur the inevitable loss of wood involved in charcoal-making (Meta, 1982, p. 5-6). A somewhat more complex substitution for wood (or charcoal) is gas, either biogas from fermentation or producer gas from the chemical treatment of biomass. From the available evidence, particularly from a fairly long record of experience in India, it appears that biogas is a reasonably economical energy substitute for wood and charcoal burning in the home. It has other advantages also: it produces less smoke, it provides a fairly good fuel for household lighting, and it concentrates the valuable fertilizing components of animal waste instead of releasing them into the air. However, biogas systems need a larger feedstock of animal wastes than is available to most households, and so they would often have to be developed on a scale larger than the household in order to have a significant impact on household community or national energy needs (Cecelski et al., 1979, pp. 89-92). But bioenergy projects that try to substitute a centralized system for a household-by-household approach tend, understandably enough, to suffer from inherent problems in organizing the system of gas distribution and in managing other aspects of the engineering (and human) systems.

Kerosene is used for lighting and cooking in many areas. For lighting purposes, fuel alcohols, especially ethanol, are a close substitute for kerosene. Alcohols could also substitute for cooking purposes, in theory—in practice the cost of alcohol, especially for cooking, is often too high.

Similarly, synthetic hydrocarbons from biomass could replace kerosene, or synthetic liquid petroleum gas (LPG)—e.g., synthetic propane—could be produced to supply a growing market for bottled gas in many areas. Here again, relevant commercial experience even in the industrialized world is miniscule, but it would be rash to assume that such synthetics would be even close to being cost competitive in the near future

Process Heat

Process heat for Third World industry (and commerce) is already partly supplied by biomass sources such as wood and charcoal. This is typically true of commercial enterprises like bakeries and of such industries as brick-making, although there is also some history of iron smelting using charcoal in Brazil, for instance (see chapter 10). Such biomass-fuel-burning installations could be a prime target for bioenergy projects designed to improve and extend wood and charcoal production and conversion.

Unfortunately, process heat in industry and commerce in the modern sectors of developing countries has more typically tended to overrely on petroleum products. Synthetic "bio-oil" would again be the most straightforward way of converting these energy end uses to biomass fuels, but the problem of untested synfuel technologies and high projected costs remains. Somewhat more plausible alternatives include the use of oils (from modernized wood distilling systems—see below in this chapter) and mixtures of finely divided charcoal and fuel oil ("char-oil slurries") (MITRE, 1981, pp. A-20 - A-31). These biofuels (or semi-biofuels) are not, however, completely interchangeable with oil in common boilers and other heating systems. Changes usually have to be made to burners, pumps, and storage facilities to compensate for the greater corrosive properties of pyrolytic oils and char-oils and higher rates of crud [9] formation in the system. Furthermore, detailed technical restrictions and heat penalties may apply in practice.[10]

Gasification is another biomass fuel technology that could either be adapted to use in existing oil-fired systems or could supply new gas-

9. "Crud" as a technical term denotes clogging deposits of detritus in pipes, for example.

10. For example, charcoal and char-oil slurries have been used in the cement industry in Brazil. Charcoal has been used for heat at the "cyclone tower," while a charcoal-oil slurry has been used to substitute for oil at the principal "blowtorch" (Meta, 1982, pp. 5-7). The substitution involves fairly serious restrictions and penalties. First, only charcoal (not wood) can be used in the slurry because of the fineness of the particles required and the low moisture requirements (Meta, 1982, pp. 5-8, 6-10). The charcoal proportion had to be restricted because of viscosity requirements, and even at that, a 38 percent charcoal content in the slurry forced an 8 percent derating of the plant because of the lower heat value of the charcoal (6100 kcal/kg versus 9800 kcal/kg for bunker oil) (Meta, 1982, pp. 5-7, 5-10).

fired process heat facilities for industry. The retrofitting of oil-fired plants to burning gas from biomass feedstocks is not an insuperable problem. Some adaptation has to be made to standard methane burners to use either the diluted methane of biogas or the carbon monoxide and hydrogen of producer gas or water gas (see the next section). But the use of water gas, for example, has been established for well over a century in industrial, commercial, and domestic applications.

Methane itself can be produced synthetically from biomass through somewhat more complex chemical processing, but the added expense is probably not necessary for local usage. Only when the gas must be transported long distances is reconversion of water gas or producer gas to a higher energy density gas such as artificial methane worthwhile (Schurr et al., 1979, p. 257).

Electricity

Thermal, that is, non-hydropower electricity is produced both by boilers generating steam—the predominant method worldwide in large-scale plants—and by internal combustion engines, usually diesel and usually in small-scale plants. So when examining projects involving bioenergy and electricity, the best biofuels for electricity generation are the very same that are suited for either industrial process heat—or for internal combustion engines—as already discussed above. However, one distinctive aspect in the electricity problem is the relatively large size of typical power plants. The scale of heat production for central electric generating plants is often significantly larger than for the average industrial process heat application: such increases in scale could favor certain bioenergy systems (e.g., direct combustion) over others. On the other side of the coin, bioenergy forms such as producer gas can be inexpensively produced from small facilities. Use of bioenergy gasifiers as a replacement for diesel fuel could therefore shift the economic balance to favor local diesel electric generators (autogeneration facilities) over standard central grids in isolated or low-load areas.[11]

Technologies

Examining feasibilities and calculating the costs of the various bioenergy technologies that convert wood or other biomass into modified biofuels—charcoal, alcohol, producer gas— tends to be a complex and

11. A comparison of the economics of producer gas-powered generators replacing conventional electrical supplies for Costa Rican rural areas reported that gas from biomass could feasibly replace present diesel-powered autogeneration. However, the economics of replacing central grid electricity was obscured by the (usual) complications of load factors and reliability. (Meta, 1982, pp. 6-15, 6-16, 6-23, 6-24)

uncertain exercise. However, there are a few things that should be said about the general opportunities and problems of candidate technologies that are likely to crop up in proposed bioenergy projects. We consider here direct combustion, pyrolysis, alcohols, biogas, and "synthetics." In particular, the relation of the technology chosen to the available feedstock is of considerable interest.

Direct Combustion

In practice, direct combustion of wood or other fairly dry biomass[12] has been well tested out, both in industrial countries and in the developing world. Lumber and paper mills commonly use wood waste for process heat, and bagasse is routinely used to supply heat for refining in the sugar industry and for the distillation of alcohol from sugar juice.

In substituting for fossil fuels, or in designing new types of direct combustion equipment to run on biomass, many of the engineering problems are relatively uncomplicated.[13] One general design problem is that biomass fuels have a lower heat content per unit volume or unit weight than fossil feedstocks, and therefore larger furnaces and larger boiler tube spacing will be needed, and other scale factors can also become a problem (MITRE, 1981, p. A-9).[14] Differences in moisture, density, and related amounts of wood sap will affect combustion, and so some kinds of wood are better than others.[15] The mineral content, or ash, also varies between different biomass feedstocks: variations in the amount of slag formed during combustion, with associated clean-up and corrosion problems, will therefore require coordination between types of feedstock and boiler design.[16]

Direct combustion can also be facilitated by manipulating the form of the fuel for more efficient combustion or transportation. Wood is often burned in modern applications, for example, in the form of chips. Briquetting or pelletizing wood or plant wastes is widely practiced to improve combustion efficiency (see Meta, 1982, passim). At

12. The biomass should have less than about "50 percent moisture"—i.e., water and dry wood (or carbohydrate) weights about equal—otherwise excessive amounts of energy may be needed to drive out the water (see MITRE, 1981, p. A-7). Note that Brazilian experience suggests that wood typically should be stored to "air dry" so as to reduce its moisture level from 60 percent to less than 30 percent (Meta, 1982, p. 3-2).

13. Pulverized charcoal, for example, is a good substitute for pulverized coal. It must, however, be kept dry, and the dry charcoal powder must be protected from accidental combustion— for example, by storage in a nitrogen atmosphere (Meta, 1982, p. 5-2).

14. There are two major classes of large-scale direct combustion, grate burners and suspension burners (BOSTID, 1981, p. 84—see their reference, Karchesy and Koch, 1979 for more information on these two options).

15. The situation is somewhat analogous for charcoal. The nature of chars can vary: chars with a low content of volatile elements are difficult to ignite and exhibit very poor flame stability. (MITRE, 1981, p. A-9)

16. For this and related reasons, the use of fluidized-bed combustion can simplify the utilization of moist and "dirty" biomass fuels (Meta, 1982, pp. 5-1, 5-2).

the other end of the scale, techniques have been promoted in Sweden for pulverizing wood so that it has the handling characteristics of powdered coal.

Direct combustion in household contexts ranges from primitive wood-burning fires to fairly sophisticated ways of using charcoal. The main problems with improving the direct combustion of simple biofuels may have to do more with matters of transport planning or with disseminating end use conversion improvements rather than combustion technology itself. That is, a good "bioenergy project" might be a program that tries to improve the market for fuelwood through upgrading transportation networks and vehicles, or that introduces more efficient stoves for household bioenergy use. The main difficulty with such measures usually lies in their cost. Costs of new stoves or of delivered fuelwood, no matter how modest, will often have to compete with very low or nonexistent cash costs for gathered fuelwood or wastes or for primitive stone or clay stoves.[17] Nevertheless, successful stove campaigns have been carried out under effective program organization, for example, in Korea.[18]

Pyrolytic Products: Charcoal, Pyrolytic Oils, Producer Gas

Pyrolysis is an ancient and well-known process under the simpler name of charcoal-making. Charcoal is made by burning wood or other biomass in the absence of a sufficient amount of air to achieve complete combustion. Modern technologies allow pyrolysis to be carried out under a wide variety of conditions of heat, pressure, and timing, and so allow a greater choice of end products. In general, products of three kinds are produced: solid charcoal ("char"), liquid "pyrolytic" oils, and gases of various kinds—especially producer gas, the mixture mentioned above of various gases, but dominated by carbon monoxide.[19] In general, dry feedstocks are more efficient for pyrolysis. Furthermore, to generate "clean" gas for some applications, charcoal has been shown to be preferable over wood as a feedstock. But in principle, almost any kind of biomass can be utilized in some pyrolytic application.

17. In fact, the cost-effectiveness of stove improvement has aroused some controversy: one reference reports "scientists have demonstrated" that when 3-stove cooking (i.e. primitive campfire-type arrangements) is done carefully, the efficiency can be as good as a well-designed chula or Lorena (improved stoves). The implication is that the poor are more careful about economizing on fuel supplies than had been supposed (BOSTID, 1982, p. 7).

18. Experience in Korea with improved "Ondol" stoves, under a Korean Forest Research Institute program, showed a 30 percent increase in energy efficiency by improving 83 percent of the fireplaces in almost 2.5 million farm households (FAO, 1982a, p. 74). (See also Chapter 12, below.)

19. Gas compositions vary according to fuel, balance of air, air and fuel, and temperature. Typical amounts of carbon monoxide and hydrogen in producer gas are reported as 20 percent carbon monoxide and 15 percent hydrogen for a wood feedstock, and 30 percent carbon monoxide and 5 percent hydrogen for charcoal. This means that producer gas, at about 4-5 megajoules per cubic meter, has about one-tenth of the heat (enthalpy) content of natural gas (40 megajoules per cubic meter) (Kjellstroem, 1981, p. 20).

The ancient method of charcoal-making is simple; the wood is placed into pits, ignited, and turf is piled over it. This process lets the liquid and gaseous products be consumed or escape unused. This procedure, however, is still the preferred method all over the Third World because of its low cost and simplicity of building and operation. Many new kinds of inexpensive but more efficient kilns have been proposed or actually constructed.[20] Practical applications in developing areas, however, have been handicapped by low feedstock (wood) costs or by the very small-scale or peripatetic nature of many charcoal making enterprises.[21]

Adapting charcoal-making or pyrolysis to producing biofuels for the modern sector is, however, another story. At somewhat higher costs and levels of technical sophistication, the efficiency of char production can be greatly improved, while storing the liquid and gaseous products for later use or using them for process heat.[22] Pyrolytic oil is produced by slowly heating biomass under high pressure and in the presence of a catalyst. To promote gas formation, the biomass is heated to higher temperatures under reduced pressure. These processes require complex equipment specific for each end product (see Ramsay, 1979a, pp. 4-8; and MITRE, 1981, pp. A-12–A-26).[23] However, such modern improvements need not imply production only on a large scale. Small airblown gasifiers can utilize a variety of feedstocks and can be used under boilers in combination with a conventional fuel backup for periods when biomass feedstocks are not available. As we noted above, they can also be used effectively with low moisture biomass feedstocks to fuel internal combustion engines—gas cleanup and cooling is required, however (OTA, 1980, II, pp. 134, 138-139). Producer gas from such small gasifiers are now being used in localized applications such as water pumping or electricity autogeneration.[24]

---

20. Some kilns considered in projects in the Philippines give an idea of the possibilities: metal-lined earth pits with side air inlets, 55-gallon metal drums with an added chimney, two interlocking steel cylinders with chimney (a portable system), and masonry block kilns with a closed chamber, loading doors, etc.—the "beehive" kiln used widely in Brazilian industry is an example of the latter (Hyman, 1982a, p. 24).

21. The problems with irrational charcoal tax structures and peripatetic production operations is illustrated for example, in Sudan, 1982, pp. 7-8.

22. One complication is that primitive batch processes for pyrolysis are exothermic (give out net heat), but for continuous-flow processes heat has to be added to maintain high production rates (BOSTID, 1981, p. 96).

23. Costs of pyrolytic oil from wastes have been estimated in one review (Ramsay, 1979a, p. 14) as from $4.50 to $8.00 per gigajoule (1975$).

Costs for gasification in one large scheme are estimated at about $2.50 per gigajoule (MITRE, 1981, p. C-403, 1980). In another estimate, costs of water gas (another useful mixture of carbon monoxide and hydrogen produced when pyrolysis steam is introduced in the interaction, see text above) is put at from $2 to $3 per gigajoule (Ramsay, 1979, p. 14).

Capital costs must also be considered: large gasification systems are expensive, several plants proposed for large scale production in the United States having been set in the billion-dollar range (See Schurr et al., 1979, p. 262, where other references are given).

24. Estimated costs for some small producer-gas-fired electrical facilities have seemed reasonable. Capital costs for small producer gas systems generating electricity, in the range from 50 to 150kw, have been quoted as from $700 to $1000 per kilowatt electric (Kristoferson and Kjellstroem, 1981, p. 1-8).

Alcohols

Ethanol and methanol are often used interchangeably as biofuels, but the usual ways of manufacturing them are quite different. Methanol is made from gas produced by heating the feedstock, ethanol by the biological process of fermentation.

Methanol is manufactured from water gas, a combination of carbon monoxide and hydrogen. The gas is made by passing producer gas through steam; the resultant gas has a higher heat content owing to the addition of hydrogen. Water gas is also called "synthesis gas" for a good reason: many synthetic compounds are in practice made from combining the carbon, hydrogen, and oxygen in synthesis gas—which during the halcyon years of cheap fossil fuels was universally made from natural gas. Methanol is one of the synthetics can be made by employing a suitable catalyst to capture and then reassemble the water gas molecules.[25] It should be emphasized that manufacturing methanol from synthesis gas is an established commercial process; indeed, the price of commercially available methanol in the United States fell in 1971 to as low as 11 cents a gallon ($1.80 per gigajoule) or the equivalent of about 20 cents for a gallon of gasoline.[26]

In modern times, however, the synthesis gas itself has long been made commercially only from natural gas, and any practical system for methanol from biomass needs to demonstrate a cost-effective method of carrying out the first step in the process —wood gasification. A recent review of Brazilian and U.S. experience estimated costs of methanol derived from bagasse or wood feedstocks as roughly competitive with gasoline prices.[27]

In principle, there is no reason that ethanol could not be made in the same way as methanol: a different catalyst could combine carbon, hydrogen, and oxygen from synthesis gas to form ethanol molecules. In practice, however, interest has centered on the age-old, long-familiar method of making alcoholic beverages like beer and wine—that is, fermentation by yeast. Fermentation has been commercial for millenia, but it has two disadvantages for present-day fuel alcohol production. One is that wine and beer—or for a proper comparison, distilled spirits like vodka—have been and still are much more valuable products per unit weight than gasoline, even at today's oil prices; one consequence is that (fermentation)[28] alcohol production has not been developed in

25. So: $CO + 2H_2 \rightarrow CH_3OH$, in the presence of copper or zinc oxide.

26. Methanol prices since then have risen considerably, at one point up to 50 cents or more per gallon (or roughly the equivalent of gasoline costing $1 and more per gallon). This increase was quite disproportionate to subsequent rises in the cost of the natural gas feedstock, ingeniously confounding all free-market predictions.

27. Costs (1975$) were estimated at between $5-10 per gigajoule (Dunkerley et al., 1981, Table 8-A-2, p. 216).

28. Alcohol for chemical and industrial purposes is usually made from ethylene, a petroleum by-product gas.

the cost-conscious environment necessary for it to compete with gasoline. The other problem is indirectly related to the first one: fermentation works only for sugars that can be directly fermented by the enzymes in yeast and for grains (starches) that can be readily changed into such fermentable sugars by another kind of natural enzyme—such as those produced by the sprouting (malting) of barley seeds. This fermentable sugar constraint is serious because sugar and grain feedstocks are both very high-priced commodities. These high feedstock costs are not only unfortunate because they make present biofuel costs high; the fact that sugar and starch feedstocks form a large proportion (50 percent or more) of present total alcohol costs means that it is very difficult to lower total prices by improving nonfeedstock technology.

Most plant material, however, consists of chains of cellulose or other forms of carbohydrates that are not as simply constructed as sugar and starch molecules. If ordinary cellulosic crop wastes or wood—much less expensive materials —could be used instead of sugar or starches, the cost of the alcohol feedstock would be much lower. Cellulose can in part be turned into fermentable sugars, but unfortunately the metamorphosis has never been easy or inexpensive. One approach is to use enzymes to break up the long molecular chains. Indeed, cows and other grazing animals employ enzymes in just this way in order to turn grass into sugar. However, the barnyard method is only economic for raising animals and the more exotic enzymes that have been used experimentally in the search for a practical cellulosic sugar-making process are so far quite expensive.[29]

Since all the enzyme does as a biological catalyst is break apart the long cellulose molecules and slightly modify the structure of the shorter molecules remaining, chemical measures can be used instead. For instance, in a process called acid hydrolysis, sulfuric acid turns cellulose into sugars. To cut costs and acid recovery problems, a "weak acid" hydrolysis process has been favored for pilot projects.[30] But this process too has some problems in keeping the sugars stable[31] as the process

29. Using an expensive enzyme has a bad positive feedback: since the enzyme is expensive, it must be recovered from the endpoint of the process, which recovery is in itself an expensive step. The wood feedstock also must be finely ground for this process, and some energy is thereby consumed (Watt, 1982, p. 183).

It must be noted, however, that technological improvements could radically alter wood sugar prospects. For example, a recent test showed that freeze-explosion pre-treatment increased cellulose reactivity so that 90 percent theoretical conversion to glucose was achieved for an alfalfa-rice straw mixture treated by enzymes (Alcohol Week, 1982, p. 8).

30. Sulfuric acid is expensive and when used in strong solutions it is customary to recover it so that it can be used again. This cost-cutting is less necessary for the "weak acid" process.

31. The sugars formed in some hydrolysis processes can be attacked by other elements in the "broth" and changed into other components before they can be recovered for fermentation purposes.

proceeds and in being able to tolerate nonuniformity[32] in the biomass feedstocks.[33]

Biogas

Biogas is generated by allowing organic matter to decay in oxygen-poor environments. It is in fact identical to the marsh gas familiar to fresh water wetlands and contains a high percentage of methane or natural gas.[34]

As might be expected with any such natural process, any type of organic matter is in principle fair game. There exist anerobic bacteria that specialize in breaking down very long organic chains of carbon, hydrogen, and oxygen—and other elements such as nitrogen—into sugars. In turn, bacteria occupying other ecological niches change these sugars into alcohols and acids and finally convert the alcohols and acids into biogas—methane, carbon dioxide, and other trace gases (Cecelski et al., 1979, p. 36).[35]

As the term "marsh gas" suggests, one of the great advantages of biogas technology is that biomass feedstocks do not have to be dried out before use, and as a matter of fact, water is necessary for the process. Little if anything else has to be added to the wastes, although supplementary amounts of nitrogen will often help optimize the production rate.[36]

The key structural feature of biogas fermentation technology is essentially just a large underground pot, made usually of concrete, containing the biomass feedstock. This digester vessel is fitted with either a fixed concrete top or a floating top made of metal that excludes air and collects the gas emitted.[37] The gas is piped off for use; after the feedstock is exhausted, it is pumped out and used as fertilizer.

---

32. As shown in MITRE, 1981, (pp. A-32–A-35), for example, wood contains not only cellulose but also other strings of carbohydrates called "hemicellulose" and a still more complex arrangement of carbohydrate groups called "lignin." Hemicellulose tends to produce "5-carbon" sugars that are not normally fermentable into alcohol, although they may have other important functions: the sugar ribose, for example, is a component of RNA. Lignin is difficult to deal with altogether and is best used, not to make alcohol, but to make charcoal, to make structural products, or as a feedstock in plastics industries.

33. The cost of producing ethanol from various biomass feedstocks has been estimated in one review at from $10 to $18 per gigajoule (1975$). (See, for example, Dunkerley et al., 1981, table 8-A-1, p. 215.) Ten dollars per gigajoule corresponds to about 80 cents per gallon wholesale.

34. Stagnant water environments tend to be oxygen-poor because dissolved air is used up by decaying organic matter faster than it enters the system. Spontaneously burning marsh gas produces the will-o'-the-wisp favored by Gothic novelists.

35. Not all the details of the process are well-understood: presumably, the first steps of anerobic fermentation could not be readily substituted for aerobic production of sugars by enzymatic or acid hydrolysis.

36. There is also some advantage to using animal wastes, or at least in combining them with vegetable wastes because the animal wastes already incorporate good sources of the right combinations of bacteria to do the job. The corresponding drawback is that the internal bovine processes have already used up a portion of the originally ingested carbohydrates.

37. A floating metal gas holder will yield gas at a constant pressure. However, an all-concrete construction has the advantage of avoiding corrosion problems. A neoprene bag set over the digester proved useful for storing the gas in concrete units in Fiji (Chan, 1981, p. 130).

As with many "natural" processes, it is not possible to ignore environmental conditions such as temperature—which can affect the efficiency or even the start-up of the bacterial processes, inhomogeneities in the feedstocks, and so on (MITRE, 1981, pp. A-4–A-5).[38] But a good deal of experience has been accumulated on different types of digesters, distinct types of feedstocks, and various patterns of energy end uses (see for example MITRE, 1981, pp. C-110ff.).

Synthetic Petroleum Products and Methane

As mentioned in the previous section, it should not be thought that the biofuel replacements for gasoline, diesel fuel, and natural gas need necessarily be somewhat exotic fuels, such as biogas or methanol. It is feasible to make "natural gas" (methane) by chemical means, and it is also possible to make gasoline or diesel oil from biomass alone—with the aid of a good deal of expensive equipment. The key questions are of costs and commercial practicability, not of theoretical feasibility.

Methane can be derived directly from organic materials by adding hydrogen at high temperatures and pressures (Schurr et al., 1979, p. 256n). Methane, like methanol, can also be produced from synthesis gas.[39] "Petroleum liquids" have been made from coal under wartime conditions in Germany and as a part of special energy-security programs in South Africa by the Fischer-Tropsch process. This type of plant reacts a synthesis gas at high temperatures and moderate pressures over a catalyst and produces both a low octane gasoline and a gas of medium heating quality.[40] However, there is great uncertainty about costs in an unforced commercial environment, and most of the technologies have not been tested on a commercial scale (Schurr et al., 1979, pp. 256-259, 516-518).

In view of these technical uncertainties and the large capital costs envisaged—even for plants based on coal feedstocks and located in the favorable technological environment of industrial countries—proposals for such facilities in developing areas should be viewed with extreme caution.

---

38. Note the following potential problems: (1) fermenta tion stops below 10 degree centigrade, (2) sludge liquidity and pH must fall within certain ranges, (3) metallic inhibitors may be a problem, (4) scum buildup can cause difficulties.

39. This methanation has been carried out on a small scale, and involves "shifting" the hydrogen-carbon balance by adding more steam to synthesis (water) gas (CO + H2), in the process turning some carbon monoxide into carbon dioxide, while combining other carbon atoms with the hydrogen to form methane (CH_4), and then removing the leftover carbon dioxide by absorption (Ramsay, 1979a, pp. 9-11).

40. Original estimates for methanation plants using the "Lurgi" process, with an output of 250 million cubic feet per day, were proposed in the early 1980s at capital costs of $1 billion or more.

An "M-mobil" process for converting (biomass-derived) methanol into gasoline has also been proposed as a possible option: however, as noted, methanol can be used directly as a motor fuel.

Planning Procedure: Conversion Technologies

The most prominent fact affecting a practical assessment of any bioenergy technology is that there exists a very large number of possible bioenergy conversion technologies. Indeed, in many cases technology decisions in the field will have to be made from a wide variety of options. However, considerable simplification in the selection process might be made by using certain lessons learned from examining the type of problems discussed in the demand, land use, and crop chapters. For example, in arid and semi-arid regions, it may be technically feasible to grow sugarcane to make alcohol—*if* irrigation water can be procured. A big *if*! In the majority of cases such an energy crop would be much too expensive for local circumstances and should be quickly passed over for more practical ideas.

If the rivers or lakes of the country contained large masses of water hyacinth, for example, then a technology for harvesting and using this aquatic weed for fuels for power plant boilers could be investigated. In practice, with such a high moisture content biomass feedstock, this might mean planning for methane fermentation (biogas) facilities.

At this point in the planning, of course, it would still be too early to make a decision about the bioenergy system, since cost trends and also environmental effects, marketing problems, and organizational difficulties have yet to be treated. Nevertheless, at this stage in the bioenergy project analysis, some of the more promising bioenergy technologies can be identified for further study as suited to the general picture of demand, land use, and crop feasibilities in a particular country.

Often several technologies could help supply such typical priority energy demands as, for example, a substitute for diesel oil for buses and semi-trailers. In this case, if the analyses described in the crop and land use chapters showed that vegetable oil crops were indeed a possibility, one would naturally investigate the technologies needed in harvesting and processing sunflowers or coconuts or whatever crop seemed practical. On the other hand, it might be that a 50 percent replacement of diesel fuel by alcohol could satisfy engine performance requirements. In that case, if there were large areas of poor, underutilized land suitable (and available) for growing, say, cassava, then harvesting and fuel alcohol industry processing, fermentation, and distilling facilities should be studied.

If the highest-priority national need were to replace oil in electricity generation, one would do well to consider different methods of combustion of biomass for steam production. Both direct combustion of wood, charcoal or sugarcane bagasse and the combustion of gas from biomass gasifiers are well-established technologies that can be used in commercial power plants.

At this stage, a general project "plausibility" should be the thing to be determined. As we shall see in the next chapters, the cost trends, questions of scale, and infrastructural and marketing constraints may complicate matters. And as the succeeding chapters will show, the total social impact of the project must be taken into account as it relates to the environment, the economy in general, and to key national planning priorities.

Conclusion

Bioenergy conversion technology is a complex, rapidly changing topic. The short review here examines the question of which technology will fit into which end use. Some technologies such as traditional wood, charcoal, and waste fuels for households appear to be somewhat restricted in scope. But the general outlook indicates a rather large choice between different options. For motor fuels, for example, both alcohols and producer gas, as well as methane from biogas can serve as substitutes for gasoline and diesel. For process heat, liquids and gases as well as woodchips and charcoal and pyrolytic oils can be used. All these can also be used to generate electricity.

The status of the technologies themselves, and in particular their appropriateness for various feedstocks, is another key question. Direct combustion of biofuels for heating boilers for process heat or electricity can make use of a wide variety of feedstocks—however, moisture and high ash content can cause problems. Similarly, charcoal and other pyrolytic products can be produced through low level technologies, although the higher technologies that produce producer gas—increasingly in small-scale decentralized operations—benefit from having low moisture feedstocks and in fact prefer the use of charcoal over wood. Biogas is especially favored for moist feedstocks, since water has to be added anyway.

All these facts, plus the nature of demand, land use, and crops, can be used in putting together a group of "plausible" in-country options.

7

Environmental Impacts and Controls For Conversion Technologies

Introduction: The Parallel With Coal and Oil

The environmental impacts of bioenergy conversion technologies cover much familiar ground, and therefore the task of the bioenergy planner is much easier than it might otherwise have been. The reason: the impacts from bioenergy conversion of carbohydrates derived from living tissues share a great many characteristics with the well-studied environmental impacts from conversion processes involving the carbon and hydrogen in coal and the hydrocarbons in oil and natural gas.

There are numerous exceptions to this parallel, of course, because the details of the composition and structure of organic compounds are significant. The products of alcohol combustion are different from those of gasoline, for example. And then there are differences of scale. Since biomass tends to have less energy per unit weight than fossil fuels, it makes sense to transport biomass feedstocks lesser distances. Therefore, it is economical for biomass facilities to be relatively small operations characterized by more localized environmental impacts. Furthermore, there exist biofuel processes such as fermentation that are totally foreign to the fossil fuel conversion spectrum and have entirely different environmental impacts.

There are, then, many similarities in combustion and conversion processes, and much of the data from the study of synthetic fuels based on coal can also be applied to wood. The planner can also take advantage of work already carried out under AID auspices reviewing the environmental effects of bioenergy conversion technologies (see MITRE, 1981).

Tradeoffs involving the environment may differ from country to country. Indeed, often there may be little concern with effluents. However, the advantages of building control measures into new bioenergy conversion facilities—versus the extra expenses of retrofitting them later—suggest that careful thought be given to the environment even in the early stages of planning.

We therefore present here a brief review, first of the environmental impacts and control possibilities for the conversion of biomass to particular biofuels, and second of the effects of biomass fuel effluents in a number of end use applications.

Conversion Technologies: Environmental Impacts

This section discusses the environmental impacts of the conversion technologies for biofuels briefly described in chapter 6. "Intermediate" energy conversion is treated here, as well as large-scale direct combustion for process heat. Other end use conversion, such as burning alcohol in motors or wood in stoves, is considered in the next section.

Direct Combustion: Boilers and Process Heat

Direct combustion of biomass will be increasingly utilized in firing large-scale boilers for both process heat and the generation of electricity. Although wood and crop residues have very low levels of inorganic ash and a low sulfur content compared to residual fuel oil—the fuel they would usually replace—they tend to generate more particulates in their emissions.[1]

The main concern here of the bioenergy planner is to see that combustion conditions are maintained as nearly complete as possible so as to minimize the volume and noxious nature of particulate organic emissions from stoves, boilers and ovens (OTA, 1980, vol. II, p. 147). Incomplete combustion may result from either insufficient air intake into the combustion area owing to poor furnace design or to an excessively high water content in the fuelwood. If moist biomass is fed into the furnace, this water must evaporate to allow the wood to ignite; boiling out the moisture tends to cause a temperature drop and less complete combustion, with an increase in particulate emissions. Some new furnace designs can make combustion more complete even in the presence of moist feedstocks.[2] Artificially force-drying the biomass first will also help, but care must be taken then to prevent high particulate emissions from the dryer itself (OTA, 1980, vol. II, p. 144 and p. 147).

1. See Appendix 7-A for definitions and a discussion of pollutants from woodsmoke.

Some types of biomass, e.g., cotton trash with 1.7 percent sulfur, have sulfur levels comparable to coal (OTA, 1980, vol. II, p. 149).

2. Adequate design of the furnace may reduce the emissions considerably, i.e., downdraft furnaces force the combustion gases back through the fuel bed, producing a secondary combustion, thus burning most polycyclic organic matters that could have resulted from incomplete combustion conditions in the higher moisture wood on top of the pile.

Impacts from large direct combustion facilities on land use and water have been noted as having strong similarities to those for coal (Schurr et al., pp. 374-378, especially p. 375). Lower ash levels, however, in general mean that the problems from wood are relatively less severe.

Depending on local environmental requirements, and on the efficiency of the furnaces involved, some sort of *control* technologies may be necessary to reduce emissions—especially of particulates. Scrubbers, electrostatic precipitators, and baghouses are established technologies for coal, but the effectiveness of particulate control technologies for large wood-fired boilers may be in doubt[3] (DOE, 1979, p. 19), and the control question requires close investigation.

Thermochemical Processes: Pyrolysis

Pyrolysis or incomplete combustion of biomass can produce a number of forms of fuel, depending on temperatures and pressures (see Chapter 6). For simplicity, the pollutants typical of low temperature pyrolysis ("charcoal-making") and high-temperature pyrolysis ("gasification") are described here. These cases can be used as a guide for studying the impacts of specific pollutant streams from a particular pyrolytic operation. Methanol, produced secondarily from a typical biomass thermochemical gas feedstock, is also treated.

Charcoal-Making: Low Temperature Pyrolysis. A good part of the emissions from charcoal ovens consist of particulates or char oils. But emissions from charcoal making include nitrogen oxide (NO_X) compounds in low percentages: about half the biomass feedstock's original nitrogen and sulfur (usually low) go into the gaseous emissions. Effluents of "polycyclic organic matter" (POMs)[4] and other organic and inorganic (ash) compounds in vapor or particulate form typically represent a full third of the total original wood mass.

Plants that are small or use primitive technologies will encounter economic difficulties in attaining the more efficient means of pollution control. The key to minimizing pyrolysis emissions of more complex pollutants is maintaining high combustion temperatures, but this goal often can only be reached by using expensive modern technology. Alternatively, if the pollutants are scrubbed out or recovered through condensers, they could be used as a substitute for fuel oil or as a chemical feedstock. However, such an "elimination" of these emis-

3. Note that baghouses (fabric filters) are the only feasible control mechanism now available for collecting particles below a few microns in size with a 99 percent or greater efficiency (OTA, 1980, vol. II, p. 145); these particles are believed to present the more serious health risks.

4. See Appendix 7-A.

sions—by designing them into the process as co-products—is often not economically feasible, especially for the more primitive forms of charcoal making.

Thus, emission control technologies or by-product operations for effluent control may only be appropriate for larger, more modern operations. Fortunately, however, the vast majority of the world's charcoal is probably made in small-scale migratory operations (usually in remote rural areas) in the vicinity of the source forest. Therefore, emissions tend to be dispersed through a large area of low population, and their adverse effects are thereby somewhat mitigated.

Gasification and Other High Temperature Pyrolysis Processes. Gasification and the production of pyrolytic liquids produce two principal kinds of environmental impacts: water effluents and air pollution. The quantity and mix of air pollutants is a function of (i) combustion and gasification conditions; (ii) environmental controls; and (iii) the chemical makeup of the feedstock[5] (OTA, vol. II, 1980, p. 148). Particulates are the typical effluent of concern but should not pose a substantial problem in well-designed systems.[6] In addition to the control techniques discussed above for charcoal, another solution is to extend the conversion process rather than burning the raw gas producer gas as is: it can be cleaned and upgraded to make pipeline gas (artificial "natural gas") or used as synthesis gas for the production of methanol. Both processes will eliminate or greatly reduce most of the more toxic pollutants in the final product.

Some of the tars produced in gasification or liquid production appear to be carcinogenic and could be a risk to in-plant workers. While there is little information to date on quantitative risks, measures to reduce occupational risk levels are available (OTA, 1980, vol. II, p. 149; DOE, 1979, p. 20). However, the major environmental problem is probably the proper off-site disposal of the water and organic liquids (and solids) produced[7] (MITRE, 1981, p. A-15). The presence of water in the gas produced will normally require drying, and disposal problems will thus be increased if the gas is to be upgraded or used for synthesis (e.g., methanol production). Attention therefore needs to be paid in general to such problems as leaching from product storage piles[8] and

5. Attention should be paid to feedstocks that may or may not be contaminated, depending on the farming and harvesting techniques used. (For example, pesticide residues may be present.)

6. Particulates from one small down-draft gasifier were put at only about 0.04 kg/GJ (output energy) (AID, 1982, Annex II, p. F-4).
There is always the possibility present of leakages of raw product gas, though the risk of this may be reduced in low pressure biomass gasifiers.

7. This is especially true for fixed-bed up-draft gasifiers in which higher percentages of organic liquids are produced (MITRE, 1981, p. A-15).

8. Leaching from by-product chars may be a problem when the char is brown char (incompletely carbonized), although black char (fully carbonized) is similar to charcoal and less polluting (OTA, 1980, vol. II, p. 149).

to the disposal and storage of process wastes and by-products[9] (OTA, 1980, vol. II, p. 149).

The production of some organic liquid effluents can be reduced by using low moisture feedstocks, or design modifications (such as the recycling of hot gas) can reduce organic liquids in the final product.

At any rate, biomass gasifiers and liquid generators should have a lesser impact on the environment than similar-sized coal processes.

Hydrocarbon Liquefaction and Methanol Synthesis. As described in chapter 6, "synthesis gas" from a biomass gasifier can be converted directly into petroleum-like liquids (hydrocarbons) or into methanol by an appropriate catalyst.

Biomass liquefaction to hydrocarbons produces gaseous effluents similar to those produced from gasification. These effluents can be controlled with technologies generally available in the petroleum refining industry.

Impacts from by-products of the methanol conversion process on the environment can occur in the form of water pollution and atmospheric emissions from the flaring of residues and contaminants. The main hazards with methanol, however, are the risks to human health of the final product itself. In particular, if it is ingested even in small doses it can cause optic nerve damage and blindness. In the industrial context, the main hazard is from vapors, where the risks exist but are not well understood in a quantitative sense.[10]

Pollution controls for biomass-based plants would be similar to those already in place in existing methanol plants using fossil feedstocks. Water treatment of phenol tars and other organic compounds from the scrubber wash can utilize conventional refinery procedures.

Biochemical Conversion Processes

Impacts on the environment from biochemical conversion processes consist primarily of water pollution, not air emissions. The reason is that pure biochemical conversion processes emit only small quantities, if any, of noxious gases but require large quantities of process water—in addition to the high water content of many biomass feedstocks. We consider here the impacts from two types of ethanol conversion processes —sugar or starch fermentation and wood hydrolysis—as well as those from biogas production.

---

9. Although the hydrocarbons in biomass gasification waste water should be more amenable to biological treatment than coal gasification hydrocarbons (they are more oxygenated), they may be produced in greater quantities and have a higher biochemical oxygen demand than those in a coal system (OTA, 1980, vol. II, p. 148-149).

10. The threshold limit value (TLV) for occupational exposure is 200 ppm (parts per million) in air for an eight-hour day with a ceiling of 250 ppm for 15 minutes (Coates et al., 1982, p. 125). (Threshold Limit Values represent occupational exposure levels that can be safely tolerated by healthy individuals at an assured exposure period of 8 hours per day, 40 hours per week.)

Substances from ethanol production posing environmental risks are (i) emissions associated with its substantial process heat requirements; (ii) wastes from the hydrolysis, fermentation, and distillation process; and (iii) effluents of toxic chemicals, especially from small facilities (OTA, 1980, vol. II, p. 173). The precise kinds of risks that result depend in large measure on the type of feedstock used: sugar, starch, or wood.

Sugar or Starch Feedstock Fermentation. The prime pollutant from the fermentation and distillation of milled sugarcane is the 12 liters of stillage[11]—the nonfermentable organic and mineral contents of the cane juice— that are produced per liter of ethanol. About 10 liters of stillage accumulate per liter of ethanol when corn or cassava is used as a feedstock.

If the stillage is mishandled, it may cause severe damage to aquatic ecosystems because its high biochemical oxygen demands will result in oxygen depletion of the water (OTA, 1980, vol. II, p. 174). The stillage can be disposed of in five ways: (i) "in natura" stillage as a fertilizer; (ii) production of methane by anerobic fermentation and captive use at the sugar mill or alcohol distillery; (iii) commercial production of single-cell protein and concentration of the stillage for use as a feed ingredient[12]; (iv) commercial production of potash ashes to be used as a fertilizer (Ribeiro and Branco, 1979, pp. 9, 12); (v) and land disposal methods like those now used in the brewing industry (OTA, 1980, vol. II, p. 174).

The simplest solution for the effluent is to use it as is for a fertilizer, but the plant should be located near farmland because liquid stillage can be most economically transported over short distances (Pimentel et al., 1982, p. 17). For this disposal procedure the composition, type of soil, and type of application system must be suitable so that the stillage is incorporated well into the soil system and is compatible with the soil chemistry.

Ethanol plants will also generate some airborne effluents, especially particulates from the burning of bagasse as a process fuel. Wastewater streams from boilers, condensate return, and wash waters will also contribute acidic, high-BOD (biochemical oxygen demand) effluents (Hira et al., 1983, pp. 206A, 212A). However, standard control measures should be economical and adequate for these problems in large plants; some efforts may have to be made in concentrating effluents or recovering byproducts in smaller facilitities.

11. The stillage, or effluent, results from the initial distillation step and is high in biochemical oxygen demand. The water requirements for large-scale plants can be substantial (MITRE, 1981, p. A-38).

12. The low economic value of raw sugarcane slop as a cattle feed makes it less desirable to recover it for use as is (OTA, 1980, vol. II, p. 174).

Wood Hydrolysis and Wood Sugar Fermentation. Wood hydrolysis—breaking down cellulose to produce ethanol from fermentable sugars—will produce gaseous emissions. Burning additional wood or the lignin residues from the process to provide heat energy for the hydrolysis, fermentation and distillation stages can produce various air pollutants, especially particulate emissions (Pimentel et al., 1982, p. 17). Acid hydrolysis plants also produce hydrochloric acid effluents from the hydrolytic stage (MITRE, 1981, p. A-34) and high emissions of unburned particulate hydrocarbons, including POMS from lignin residue drying and the charcoaling of the solid residues. Enzymatic hydrolysis, using acid pretreatment, will also produce inorganic salts and a liquid effluent that will have to be treated to reduce BOD (MITRE, 1981, p. A-34).[13]

By far the greatest pollution problem is the stillage waste remaining after the distillation process, especially that resulting from the hydrolysis and fermentation steps, which can account for 60 percent of the total pollution output (Ribeiro and Branco, 1979, p. 7). Problems and possibilities are similar for wood-derived stillage and stillage from sugar and starch feedstocks. Minor solid waste problems can also be controlled by recycling.[14]

Biogas (Anerobic Digestion)

As in alcohol fuels production, the biogas anerobic digestion process produces effluents (Pimentel et al., 1982, p. 18). But in sharp contrast to alcohol, biogas production can be thought of as reducing water pollution because the feedstock is thought of as an effluent itself. The digestion process turns the organic feedstock substances into the clean gases methane and carbon dioxide, plus traces of water vapor, oxygen and hydrogen and negligible amounts of nitrogen oxides (NOx) and hydrocarbons (HC). Indeed, anerobic digestion is a pollution control technology well known in urban raw sewage treatment (DOE, 1979, p. 21).[15] It tends to kill disease-causing viruses and bacteria, reducing

13. Other air emissions include fugitive dust from the raw material and by-product handling and emissions of organic vapors from the distillate process. These emissions can be controlled by water scrubbing (for organics) and by cyclones (for dust). In addition, the use of toxic or flammable reagents such as cyclohexane in the final distillation step to produce the anhydrous ("water-free") alcohol required in gasohol mixtures can pose occupational risks (OTA, 1980, vol. II, pp. 173-174).

14. There is considerable solid volume of neutralized salt residue from the acid hydrolysis process. These salts can in principle be recycled back in agricultural use to reduce soil acidity, thus eliminating disposal problems.

15. However, in addition to the contents of the effluent in rural digesters, urban sludge digesters may contain high concentrations of inorganic salts and a variable amount of toxic metals, such as boron, copper, and iron (OTA, 1980, vol. II, p. 196). Urban sludge not usable as fertilizer can be disposed of in several ways: (i) dumped onto fields, although this may result in high salt content and metal concentrations so that the land will have to be rotated; (ii) evaporated, especially appropriate in arid climates; (iii) discharged directly into waterways, or (iv) discharged into public sewage systems (OTA, 1980, vol. II, p. 196).

98 percent of the raw waste into a more benign "sludge" that retains nutrients that make it valuable as a natural fertilizer (OTA, 1980, vol. II, p. 195). Best results are obtained if the sludge is plowed back into the soil, reducing the loss of nitrogen (Pimentel et al., 1982, p. 18). It can also be used as a cattle feed as long as the effluent's pesticide content is not too high (OTA, 1980, vol. II, p. 197).

In practice, however, if retention time is not long enough, the sludge may still contain a significant percentage of organic matter that can impact on water quality. In addition, digesters can produce some ammonia, and imperfectly digested slurry exposed to air usually starts aerobic bacterial digestion, producing noxious odors.

For some applications in stoves or boilers, the gas may be scrubbed before combustion to reduce hydrogen sulfide (H_2S) corrosion and air pollution problems. H_2S scrubbing is easily accomplished by percolating the raw gas through a bed of wood chips and iron oxides (KVIC, 1978, p. 7).

Without scrubbing, venting of raw gas can cause odor problems, although the odor of hydrogen sulfide can give a useful warning of leaks (OTA, 1980, vol. II, p. 197). Methane is also explosive when mixed with air, but its flammability range is very narrow—methane/air mixtures below 4 percent or over 16 percent will not ignite. Its lower density and higher combustion temperature make it generally safer than, say, gasoline to handle (DOE, 1982, pp. 7-18 and 7-19).

Efforts may have to be made to prevent the effluent waste water from infiltrating the ground water system and degrading the water supply, especially where soils are porous and cannot naturally purify the effluent (DOE, 1979, p. 21; MITRE, 1981, p. A-4).

End Uses: Environmental Impacts

End uses also have environmental impacts. Potential pollutants from wood, charcoal, producer gas, alcohols, and biogas must be considered by the planner.

Fuelwood

Wood as an end-use fuel is utilized in a number of different forms, such as logs, chips, and pellets. However different the equipment for each of these forms, the kind and volume of emissions is mostly dependent on the size of the stove, oven, or furnace, the water content of the wood, the combustion temperature, and the mode of operation. Indeed, incomplete combustion of wood in small residential ovens and stoves can produce some of the most environmentally serious impacts as well as some of the most difficult to control—because of the inherent difficulty of regulating household activities. Two major environmental risks are associated with burning woody biomass in these small units: (i) direct health risks from lung irritants and carcinogens to residential or commercial fuelwood users, and (ii) toxic air pollution impacts, especially lung irritation, on the general population in the local vicinity (Coates et al., 1982, p. 5). Appendix 7-A describes pollutants in household woodsmoke.

Charcoal

Charcoal usually produces lower POM emissions than wood because its lower water content makes for more complete combustion. However, the fine charcoal dust formed during handling and transportation tends to make particulate emissions higher.

As we have already discussed, the ability of the boiler to maintain nearly complete combustion conditions will affect particulate emissions: hydrocarbon and carbon monoxide (CO) emissions may also decrease somewhat when combustion temperatures are raised.

Gas From Pyrolysis: Producer Gas and Its Relatives

The cleanup procedures used for producer gas vary considerably depending on the equipment it is to fuel. For gasoline or diesel motors for electric power or for automotive vehicles, the gas is generally produced on-site or on-board, and must be made free of particulates through filters and scrubbers as it enters the motor. Since the fuel is mainly carbon monoxide, which is burned to nonpolluting carbon dioxide, it produces only a small fraction of the levels of pollution of corresponding gasoline or diesel fuels. Nitrogen oxides are an inevitable result of combustion, here as elsewhere, but use of producer gas in a 70:30 mixture with diesel, for example, will lower temperatures by some 140° and therefore lower NO_X emissions (AID, 1982, Annex 2, p. F-7).

Producer gas—or its hydrogen-rich relative, water gas— can also be used to fuel industrial boilers, ovens, and other process heat needs. Direct coupling of the gasifier to the combustion equipment can eliminate the need for scrubbers. On the other hand, if its heat value is to be raised to pipeline gas specifications, the gas has to be filtered and scrubbed in the upgrading process. In either case, the final combustion tends toward completeness, and noxious emissions are lower than, for example, those from equivalent rated oil-fired boilers.

Alcohol Fuels

Ethanol and methanol-based fuels—both blends and neat (pure) alcohol—have significant impacts associated with their distribution, storage, and comsumption. The major environmental concern for alcohol blends is the air quality impact from motor and evaporative emissions, both during distribution of alcohol blends and at the motor exhaust. It must be assumed that no additional anti-emission measures will be taken beyond those already used for gasoline alone: under these circumstances, some pollutants will be somewhat less (than for pure gasoline), others more, with inevitable increases in aldehydes and other alcohol combustion products (Ramsay, 1980b, p. 39).

The direct toxicity of alcohol and gasoline blends is of less concern. Toxic impacts depend on the means by which exposure occurs: inhalation, absorption through the skin, or ingestion. Any health impacts are related to the likelihood and extent of exposure. They are, logically, less for the average consumer than for industrial workers. The toxicity of alcohol-gasoline blends is similar to that of gasoline alone (OTA, 1980, Vol. II, p. 211). Preliminary experiments on alcohol-diesel blends on diesel motor emissions indicate a reduction of sulfur oxides (SO_X), hydrocarbons (HC), and particulates.

For neat fuels—in contrast—toxicity, especially from methanol, can be the impact of greatest concern. There is no known way of fighting the problem except through careful handling. Pure ethanol is well known to be toxic, but serious illness or death from oral ingestion occurs only at very high doses. Moreover, extended exposure to high concentrations of ethanol vapors can be tolerated without serious long-term effects (DOE, 1980, p. 50). Methanol exposure, in contrast, readily leads to blindness, liver disorder, or death. Exposure through inhalation can be tolerated for up to 30 minutes at concentrations of 25 thousand parts per million without incurring any irreversible health effects. (DOE, 1980, p. 50). However, very small quantities (several cubic centimeters) have been known to cause blindness when swallowed, and numerous cases of blindness caused by absorbing methanol (used instead of rubbing alcohol, i.e., propanol) through the skin have been observed in infants (Ramsay, 1980b, pp. 18-19). Indeed, the potential toxicity impacts of a methanol fuel economy, including the possibility of widespread inhalation and skin absorption at future methanol service stations, must be considered a serious concern for bioenergy planners.

Both ethanol and methanol are relatively clean burning as neat fuels, so there are neither significant particulate matter emissions, nor polycyclic aromatic hydrocarbons, nor sulfur dioxide. They yield 50 percent less nitrogen oxides than gasoline but about the same amount of carbon monoxide, while hydrocarbon emissions are virtually nonexistent (Coates et al., 1982, p. 125, and DOE, 1980, pp. 48-49). Unburned fuel emissions do increase: these emissions are composed of methanol or ethanol and aldehydes. These fuels and partial combustion products are less photochemically reactive than unburned hydrocarbon compounds and tend to produce less photochemical smog (Coates et al., 1982, p. 125). However, the long-term effects of formaldehyde emissions, in particular, must be of concern to the planner.[16]

Biogas

Methane, the major constituent of biogas, is considered a clean burning fuel. Nevertheless, when gas is burned on site without scrubbing out pollutants, combustion will oxidize contaminants to compounds like sulfur and nitrogen oxides. In particular, hydrogen sulfide will form sulfurous and sulfuric acids—corrosive to metal—and biogas in some uses might therefore increase the cost of maintenance if it is not scrubbed beforehand. Digester gas scrubbing methods are simple and inexpensive: for example, an "iron sponge" of ferric oxide and wood shavings in which the gas reacts with iron to form ferric sulfide.

Methane from biogas as a motor fuel reduces emissions of hydrocarbons, CO, NO_X, and SO_X by a large factor over gasoline. In diesel motors, it reduces particulate emissions by a factor larger than four, but owing to lean combustion limits at low rpm combustion is incomplete and fuel combustion emissions are high. But unburnt methane, the main constituent of these emissions, is readily degradable in the course of normal atmospheric processes. For a wide vehicle operating range with biogas, the gas must be compressed and stored in steel cylinders at 3500 psi, causing a number of safety risks involving fire and explosions.[17]

Methane is nontoxic so there are no threshold limit values (TLV) for repeated exposure and it constitutes no health hazard for skin

---

16. However, 3-way oxidation catalysts have been effective in sharply reducing aldehyde levels (Coates, 1982a, p. 125).

17. Recent reports on CNG motor safety show, however, no CNG-related deaths (AGA, 1983, p. 19). The smaller fire hazard CNG poses in comparison to gasoline fueled motors is related to two major factors: (DOE, 1980, pp. 7-17 and 7-18).

(1) The thick steel walls of high pressure cylinders and fixtures are much more accident-resistant than the thin-walled gasoline tanks.

(2) Physico-chemical properties of methane also help reduce fire and explosion hazards in relation to gasoline—methane has half the density of air so when leaks occur, the gas rapidly dissipates. In addition to its ignition temperature being 50 percent higher than gasoline, it has a very small air to fuel ratio window in which it can burn. However, ignition temperature in relation to gasoline and diesel fuel alone does not measure the relative fire hazard of these fuels because relatively weak thermal ignition sources can ignite gaseous but not liquid fuels; some experimental uncertainty therefore exists about methane explosions.

contact. As to inhalation, it is a simple asphyxiant—it smothers by crowding out oxygen—but nontoxic. Effects of accidental fuel release upon terrestrial or aquatic organisms are nil, and air emissions of methane are relatively benign as to environmental and health effects. If biogas is liquified, the liquid can cause severe damage to skin and materials due to the cryogenic temperatures involved (−162°C), but when released, the liquified methane boils to gas and disperses rapidly (DOE, 1982, pp. 7-18 and 7-19).

Planning Procedure: Environmental Controls On Conversion

Environmental impacts from bioenergy conversion facilities need not rule out a plausible bioenergy option in most cases. Environmental controls are readily available for many of the modern combustion and thermo-chemical options; often technology from the petroleum industry is readily available for all these technologies. While environmental controls are often at present neglected, especially for primitive technologies, there are two disadvantages to this neglect for modern bioenergy projects. First, environmental concerns may be expected to grow in the Third World as the consequences of pollution become more and more apparent. Second, in many cases environmental control goes along with efficiency. This is especially true for direct combustion and gasification, where careful control of combustion temperatures will often save energy as well as reducing effluents.

In biochemical conversions such as fermentation, the value of thinking ahead is emphasized by the title of a recent paper, "Ethanol Stillage: A Resource Disguised As A Nuisance" (Ribeiro and Branco, 1979), where a part of the process stream that is an effluent or pollutant when discharged into a nearby stream can become a fertilizer when applied to local soils.

The environmental effects of some technologies, however, could be considered as influencing the choice of options. For example, the choice of the methanol option may be inadvisable until such time as the questions of general methanol toxicity have been more fully investigated in the context of a society where large quantities of methanol are apt to be absorbed through the skin, mouth, or lungs. Furthermore, some options may appear more favorable when contrasted to other options having relatively large environmental impacts. In view of the environmental disadvantages of woodsmoke from many kinds of simple stoves, for example, the biogas option may be given extra credit for its environmental advantage—clean, smokeless burning.

Conclusion

Environmental impacts from bioenergy conversion facilities can be complex. They can involve serious health risks and require expensive control measures. It is some consolation, however, that bioenergy does not generally surpass—or usually even equal—the impacts from oil or from coal-based synthetics or other established fuels. In fact, probably the worst environmental problem in the conversion area is that of smoke from biofueled stoves—an age-old problem that may yet stimulate more aggressive environmental policy action.

APPENDIX 7-A

Pollutants in Household Woodsmoke

There are several components to woodsmoke emissions. These emissions occur during all kinds of combustion, but are particularly significant under the conditions of low temperatures and pressures characteristic of household use:

1. Particulates, a catch-all term, includes inorganic ash, condensable organics, and fine carbon char. About 50-75 percent of the particulates are respirable and represent the greatest health hazard from wood burning. The rate of emission is a function of several factors: wood moisture content, wood species, furnace design, and speed and completeness of combustion (Coates et al., 1982, p. 238). Particulates are a worrisome component, especially during temperature inversion conditions. Condensable organics form about two-thirds of the particulate matter formed by residential wood combustion units (OTA, 1980, vol. II, p. 143).
2. Hydrocarbons are released under poor combustion. Sunlight converts them into peroxy compounds—strong oxidizing agents which can oxidize nitric oxide to nitrogen dioxide. This cycle displaces the normal pathway for ozone reduction with a net result of localized increases in the ozone concentration characteristic of photochemical smog (Coates et al., 1982, p. 138).
3. Polycyclic Organic Matters (POM), also known as Polycyclic Aromatic Hydrocarbons (PAH), may present the greatest health risk from woodsmoke, as they bear carcinogens and carcinogenic-like material (Coates et al., 1982, p. 142 and p. 145).[18] POMs are emitted by all combustion sources: they are products of incomplete combustion and may increase with the use of air-tight stoves. POM emissions from wood stoves can be greater on a per BTU basis than fossil-fueled stoves. POM risks are notable because (i) they are relatively likely to reach human tissues when they condense into particles in the flue gas, especially smaller particles

18. Carcinogenics such as benzo(a)pyrene, benzofluoranthene and acetaphthylene are emitted; emissions of benzo(a)pyrene are estimated to be 50 times greater than emissions from oil-fired household facilities (Coates et al., 1982, p. 138). These carcinogens make up as much as 5 percent of the POM from typical combustion.

that are less likely to be captured by particulate controls and more likely to penetrate the lungs; (ii) several POM compounds are potent animal carcinogens and suspected contributors to lung cancer and (iii) they are suspected of causing or contributing to an added incidence of chronic emphysema and asthma (OTA, 1980, vol. II, pp. 143-144).

4. Nitrogen oxides (NO_X) are produced in small amounts (Coates, 1982, p. 138). Because of the lower heating value of wood and because higher combustion temperatures are not reached, reducing the formation of undesirable nitrogen oxides (MITRE, 1981, p. A9), biomass NO_X emission levels are slightly below those from oil-fueled boilers and stoves (Coates et. al., 1982, p. 140).
5. Sulfur oxide (SO_X) emissions are relatively unimportant (Coates, 1982, p. 138) because wood sulfur levels are equal to or less than 0.2 percent, and they are one order of magnitude smaller than oil-fired stoves and residential space-heating furnaces.
6. Carbon monoxide (CO) levels are relatively high, as are unburned hydrocarbons, in emissions from small wood burning units because of low combustion efficiency (OTA, 1980, vol. II, p. 143). These emissions may be limited in scope to the immediate "downstream" effect of a residential unit in rural areas. But a concentration of these household units can pose serious emissions dispersion problems in populated areas subject to inversion phenomena.[19]

19. This chapter is based on drafts by Odwaldo Bueno Netto, Jr. and earlier drafts by Richard Kessler.

8

The Outlook for Costs and Financing and the Implications of Scale

Introduction: Money and an Uncertain Future

Bioenergy planners will normally be intimately acquainted with the costs of specific projects they are evaluating. Even when they are searching for new projects to undertake, planners will find that most engineering descriptions come supplied with adequate cost analyses. But the possession of normal cost accounting figures does not ensure that all problems of cost can be laid to rest. What, for example, will be the costs in future years? What are the costs and prices of alternative fuels—which, after all, will determine whether the bioenergy project makes economic sense? Will financing be available for a project, or if it is available for a single project, what about financing for a larger multi-project program? Finally, the thorny questions of foreign exchange, inflation, and the changing terms of trade for developing countries can interfere with the best-laid bioenergy plans.

Another elusive factor is the question of scale of size of facilities. This question is not merely an economic problem, but also has social dimensions. In bioenergy, as elsewhere, "bigness" has implications for the development process and the kind of society it builds.

Cost Trends for Factors of Production

Typical costs of bioenergy cultivation and conversion technologies have been reviewed in various sources (see for example, Mitre, 1981). Actual costs, however, are very site-specific. Planners therefore will have to judge for themselves the reliability of the cost data presented to them for a particular project in a particular place.

A more general and interesting question is that of cost escalation, that is, the general outlook for future factor costs in the developing countries. The trend of costs will be dependent on the rising (or falling) costs of land, labor and management, and capital. In a theoretical sense, this problem is not as complicated as it might be, because only the cost of these factors relative to other costs in the economy is important; as development proceeds, GNP should rise and so should factor costs in general. The bioenergy planner, on the other hand, is only interested in bioenergy vis a' vis other technologies. Firm conclusions are still hard to come by, but there are some trends that can be watched for.

Land

According to elementary economic models, since land is limited as a factor by its very nature, the price of land could be expected to rise relative to other costs as development proceeds and demand increases. This cliche should be a reliable guide for countries like Egypt, where arable land is very limited compared to population, but less so for countries like Brazil, where large unused tracts of land can be brought into cultivation through such infrastructure improvements as new road networks.

Bioenergy economics depends strongly on land costs. One can expect that bioenergy will therefore incur higher costs relative to alternatives if the cost of land rises relative to other factors. Large bioenergy programs will themselves contribute to this land pressure. It has been estimated that the fuel alcohol program in Brazil, for example, will have taken over about 2 million hectares of new land by the end of the period 1980-1985. In the Brazilian case, this contribution will add to land pressures from other sources—export crop expansion, for example, brought 3.5 million hectares of new land into cultivation during the period 1966-1968 through 1976-1978 (Poole, 1983, p. 34). Often, new land is not brought into cultivation, but subsistence food crops or export food and fiber crops are displaced. Such a phenomenon has been noticed in Brazil and has also been calculated as a probable outcome in one model for Costa Rica (Celis et al., 1982, p. 48).

A revolutionary possibility for future land costs is that controls might be placed on land prices if bioenergy programs come to be treated as priority items by national governments. In one theoretical analysis, for example, it was calculated that large fuel alcohol programs might not be able to meet established quotas within financial limits unless land prices were controlled (Ramsay and Jankowski, 1982, p. 8).

Labor

Does the cost of labor go up as countries become industrialized? There is some evidence from Korea: even though working-age population as a percentage of total population increased from 55.4 percent to 64.9 percent (a 17.1 percent increase) during the period 1964-1978, participation in the labor force (employment level) increased from 54.5 percent to 58 percent (only a 6.4 percent change) over the same period. This trend would imply more competition for labor and higher wages, although some evidence of labor shortages had been reported (Korea, 1981, pp. 28-29). Also, during the period 1963-1974, labor productivity averaged over all sectors increased at an annual rate of 6.0 percent (Korea, 1981, p. 30), and wage rates did not always reflect such increases in productivity, especially in those cases in which there severe shortages of labor. Therefore, if the Korean experience is applicable elsewhere, one should not expect bioenergy schemes to be threatened with overall increases in wages (labor costs) that are totally out of line with productivity increases.

Bioenergy has one special concern that is often very different from urban industrial employment in a political and economic sense—the rural labor question. Bioenergy costs may be relatively low if there is a supply of underutilized, inexpensive labor in areas where biomass crops are cultivated or converted. Evidence on such underutilization of labor is mixed, however, and the question is controversial. Furthermore, bioenergy can itself contribute to rural labor pressures. In one economic model calculation for Costa Rica, a program comprising four large distilleries affected the seasonal labor demand in one region of the country in a radical manner, doubling the peak need for labor at harvest time (Celis et al., 1982, p. 55).

The supply of management, at least competent management, is fully as short in developing as in industrial countries. Because it has been postulated that low labor productivity is most typically the result of poor management (Hirschman, 1958, p. 146), rises in costs of management (i.e., shortages of competence) will be doubly important. If this management-labor linkage holds in the case of Korea, for example, one could infer that the relatively high rate of growth for Korean labor productivity implies a high "productivity" for Korean management.

Capital

The cost of capital, as measured by the cost of borrowing money—the interest rate—is a complex story. Although "real" rates—adjusted for inflation—for borrowing private money often ran at levels of 20 percent or so in the period immediately preceding 1975, interest rates from governments ranged from 8 percent to -16 percent in real terms (Dunkerley et al., 1981, p. 224). These figures probably reflect the low real rates over that decade of loans from international banking institutions. Even for the industrial countries, it has been noted that real interest rates in recent decades have been very low and that the evidence on the supply side does not point to future sharp discontinuities in capital availability (Gordon, 1979, pp. 40-41).

As far as the productivity of capital—returns on investment—is concerned, it has been noted that the productivity averaged over developing countries has been higher than the overall figure for industrial countries (Gordon, 1979, p. 39). This corresponds with the usual economic wisdom that capital is more productive where it is scarce. It is difficult to draw any definitive conclusions from this isolated observation, however.

Costs of Alternatives

The financial viability of a bioenergy project depends on the costs of alternative fuel sources: cheap oil will be preferred to expensive producer gas. The relevant fuels are customarily oil or hydroelectric power, but coal and other fossil fuels must also be considered, as well as newer renewables such as solar and wind generators. Future costs must be estimated; in many cases these can be only inferred indirectly through an assessment of the long-term supply (resource) outlook. Competition among different biomass sources of energy must also be considered a factor.

Oil

Whether a nation is an oil importer or exporter, the economic cost of oil as a competitor to biomass will be determined by the world market price for crude. The nightmarish uncertainties of attempting to predict prices of petroleum over the next decade do not need to be overemphasized. Economic factors such as GNP growth rates combine with political factors —such as wars and revolutions and the impact of oil earnings, or lack of them, on the political stability of the exporting countries—to make the prediction game very difficult indeed. It can be noted that analysts showing optimism—from the point of view of oil importers—feel that prices will remain low in relation to 1979–1980 levels for some time. For example, one estimate shows prices rising to their pre-slump level ($34 official price for Saudi Light) in current dollars by 1986, with the price keeping pace with inflation thereafter. In effect, oil prices are expected in this analysis to stabilize at $3 below pre-slump levels in constant 1973 dollars (i.e. $13 per barrel) (Netschert, 1983).

Another source sees crude petroleum prices rising slightly to $31 by 1985 with a 2.1 percent annual growth in demand thereafter (Data Resources, Inc., 1983, p. 1.10). These estimates are roughly consistent with others—which for example show crude oil prices in 1982 dollars remaining below the 1981 level through 1995, registering only marginal increases between 1984 and 1985 (GRI, 1983, p. 21).

Despite these comforting projections, there are minority views that predict significant rises in oil prices over the next decade (Wilkinson, 1984). And in the event of new problems in the Near East, it is likely that planners in general and bioenergy planners in particular will be making agonizing reassessments.

Coal, Other Fossil Fuels, and Nuclear Power

The case of coal is complicated by the fact that coal contains a relatively low amount of energy per unit weight. This means that the difference between the price of domestic and foreign coal supplies can be very significant. The opportunity cost of oil—not necessarily the domestic price, which is often subsidized—must correspond relatively closely to world markets because transport costs are a relatively small part of the final price. But the transport cost of coal is so significant that each individual case must be given separate analysis.

Long-term trends have been estimated for prices of coal in the United States. One analysis shows the price of coal in current dollars as increasing from $32.23 a ton in 1981, to $42.50 a ton in 1985, to $107.10 a ton in 1995—i.e., increasing at an average rate of about 8.6 percent between 1982 and the year 2000 (GRI, 1983, p. 19). Because the United States is one of the leading world coal producers, it is therefore one of the prime prospective exporters of coal, so these figures are of some interest. Other countries such as South Africa, Poland, and Columbia, however, will have different cost structures that will determine the prices of coal from those particular countries. Despite revived interest in the coal option in recent years, the world market remains rather fragmented, with varying contractual arrangements discussed between the small group of exporter nations and importing countries (Dunkerley et al., 1981 pp. 153-154).[1]

Both long-term costs of imported coal and prospects for new domestic supplies hinge on the course of new exploration efforts for this resource. Due to a general lack of interest in coal over the past 50 years or so, exploration has been neglected: large bodies of new coal resources could well exist. For example, estimates of worldwide resources made 10-15 years ago appear to have underestimated reserves in countries

---

1. For bioenergy planners costing out this alternative fuel to assess its competitiveness, one of the most important factors is the mode of transport and the distance from source to market. Shipping costs, for example, tend to be very dependent on form of transport, with water (ocean) transport quoted at $0.0008 per metric ton-kilometer (1975 dollars), but with costs for coal transport by railroad put at $0.015 per metric ton-kilometer. This means that a 530-kilometer train trip will add approximately the same cost ($8) to a ton of coal as a 10 kilometer ocean voyage.

like Zimbabwe, Swaziland, Botswana, and Nigeria. The traditional scarcity of evidence for coal resources in many tropical countries could have a basis in the physical processes of coal formation. On balance, however, it seems likely that coal reserves in the world overall are greater than had long been thought (Dunkerley et al, 1981, pp. 140-145). Therefore, one can forsee that coal will over the medium-term future represent a relatively inexpensive fuel.

Reserves of natural gas form perhaps one of the greatest resource mysteries in the present situation. The outlook for gas associated with oil fields is probably rather dim, for the same general reasons that lie behind the lack of (long-term) optimism for oil in general. The question of "unassociated gas" may pose a different problem entirely, however. If minority theories that postulate that natural gas has an inorganic origin are correct, the total gas reserves in the world could be very large.

World shale oil reserves are fairly significant in size, on a global basis corresponding to about half a trillion toe (Dunkerley et al, 1981, p. 147). Only about 5 percent of this is recoverable under present economic conditions. Even so, it seems likely that—at the relatively low level of exploration up to now—this shale resource has also been underestimated.

It also seems almost certain that total amounts of recoverable uranium are relatively large, at least if future uranium prices rise to a sufficiently high level (Dunkerley et al., 1981, pp. 147-152). And the contribution of nuclear fuel costs to the total costs of nuclear electricity would probably be less than half even if prices of uranium rose as high as $520 per kilogram, a figure far above today's market (Dunkerley et al., 1981, p. 150). For uranium, however, one must consider the total energy package, since the only relevant use for uranium is a fuel for reactors for nuclear power plants. The capital costs of power plants are therefore a key parameter. Based on U.S. experience, the capital costs of nuclear power plants average about 33 percent above the costs of coal plants—while the costs of coal-fired plants probably average another 50 percent over the cost of oil. Even so, figures are valid only for the United States (Dunkerley et al., 1981, p. 185), and relative cost ratios in developing nations may be higher. Furthermore, such comparisons do not take into account the special problems (safety, proliferation) of nuclear power that might inhibit its development as an alternative in the Third World (Dunkerley et al., 1981, p. 191).

Bioenergy planning may have to take notice of such alternatives—particularly imported coal—but the analysis required will often go beyond the usual project scope and involve overall national energy policy.

Hydroelectric Power

The potential for new hydropower in many of the world's developing countries is enormous. Very rough estimates made several years ago show that the capacity already on line in developing countries is probably about 15 percent of the capacity that was rated as "installable" by the 1974 World Energy Conference (see Dunkerley et al., 1981, p. 162). Hydroelectric planning, however, is an art unto itself, and it is not anticipated that the average bioenergy planner will have to engage in a detailed assessment of future electricity generation from hydroelectric power. As far as estimating the costs of hydroelectric power as a potential competitor to bioenergy, however, it is probably a good rule of thumb to suggest that the costs (per unit of energy generated) of new hydroelectric installations will be somewhat below that of present oil-fired electricity generators.[2]

Other Renewable Competitors

There are a wide variety of renewable energy sources such as solar, wind, tidal, and geothermal power, that could replace present fossil fuels and might be competitive with bioenergy sources. As far as electricity generation is concerned, probably only geothermal power is of practical importance for central station generating facilities (Schurr et al., 1979, chapter 10). Photovoltaic cells and in some cases wind generators could also be useful in isolated locations far from connections with the central electric grid.

Other renewable sources compete with bioenergy in specific instances. For example, there has been some movement toward the use of solar energy for crop drying and other process heat applications in industry. But the most promising use of solar energy in developing countries to date has been solar hot water heating. In Kenya it seems to compete with other sources because of the high local insolation (solar radiation) (Schipper et al., 1983, pp. 40-43). A possible competitor of bioenergy for water pumping could be wind power. This competition is particularly formidable in cases where the schedule of water pumping could be changed without damage to crops, thus taking advantage of the sporadic nature of the wind resource without the need for expensive storage devices.

Competition Within Biomass Energy

Some bioenergy projects, such as fuelwood plantations, directly increase supplies of traditional biomass energy. Other, newer biofuels

2. This suggestion supposes that some institutional inertia is required to acquire the necessary capital and planning to build new hydroelectric facilities that only became economically feasible after the prices of competing petroleum fuels rose in 1973-1974 and 1979-1980.

can replace traditional bioenergy forms such as wood and charcoal if their costs can be made competitive. A well-known example is the biogas ("gobar gas") program in India, where animal wastes that might have been directly burned as fuel are used instead as a feedstock for gas production. A special problem that must be overcome in this interfuel competition is the nonmarket quality of much of the "trade" in traditional fuels. In the Indian case, this is partly handled by subsidizing the construction of the biogas digesters, and partly through counting the economic benefit from the slurry fertilizer by-product and the "superior" quality of the gas as a fuel. In general, bioenergy planners must be able to devise appropriate schemes for dealing with noncash competition from wood and waste fuel alternatives.

Pricing Policies

Prices and costs are not necessarily the same. Oil product prices, for example, will have to include extra transportation and distribution charges above the usual world prices before reaching the average consumer in the developing country. But even including such extra costs in delivering the product to the final end user, many governments regulate ("administer") the prices of the petroleum products and other fuels, especially electricity.

Petroleum Products

Petroleum products are often sold at subsidized prices in oil exporting countries. The costs of such subsidies are essentially invisible to the consumer and therefore the political pressures may make oil product prices too low to allow any kind of competition from modern biofuels. Some oil exporters like Ecuador, however, have gradually moved away from such policies, and others may follow, especially countries where known reserves will soon be exhausted.

Petroleum product prices are often subsidized in oil importing countries also, sometimes for more specific policy reasons. The most common kind of subsidy is one given to allow the sale of kerosene at relatively low prices. The rationale for such a pricing policy can be that kerosene should be an integral part of general program of subsidies for subsistence items—on the same basis as rice or cooking oil. Or it can be as a device to dissuade people from the use of wood and wastes and so to help conserve forest and fertilizer resources. The policy of selling kerosene at a low price was apparently pursued for both reasons in India until sometime after the first oil crisis. Since then, however, the price of kerosene has been raised to a value closer to the world-equivalent value. Indonesia is an example of an oil-exporting country where this policy has also been followed: the price of kerosene has been kept purposely low with the express intent of discouraging the cutting of trees for fuel.

Another popular administered pricing policy has been to bias the prices of diesel fuel and gasoline, two closely related petroleum products. By lowering the price of diesel and raising the price of gasoline one can discourage passenger car use and encourage freight hauling by diesel semis—given the not entirely justified assumption that more direct productive benefits come from the latter activity. This policy is followed in Brazil, where it has been tied in with the national fuel alcohol program, as well as in Costa Rica and in Haiti. In Haiti the government has reduced existing taxes (or margins) on imported petroleum products in order to shield consumers from the higher oil prices of the 1970s and 1980s; the price of gasoline, however, has been kept purposely high for "conservation reasons" (World Bank, 1982g). Such high prices for gasoline may seem to pose an opportunity for the bioenergy planner. But one must also realize that such high gasoline prices also lead to shifts to a greater use of diesel vehicles, even for passenger transportation. This leaves diesel as the prime competitor for alcohol—and diesel equipment is more difficult to adapt to fuel alcohol than gasoline engines.

Electricity and Other Fuels

Electricity pricing is also another area where discriminatory rates could pose a problem to the bioenergy planner if bioenergy *replaces* electricity in an end use. If biomass energy is used, however, as a substitute *source* of generating fuel for electricity—the most probable option with present-day technology—it presumably would participate in the same kind of rate setting and block structures as other fuels and would not suffer any competitive disadvantage.

Other fuels can also either be subsidized or discriminated against. In Korea, for example, coal is generally priced below market cost, while gas produced from petroleum has been sold at $0.41 per cubic meter, which is less than the estimated marginal cost of operations (Korea, 1981, pp. 35, 36, 40).

Capital Availability

Bioenergy planning is useless if capital is not available to implement programs. The availability of capital at the local level is a complex problem in itself, touching on the kind of social infrastructure that is necessary for many development projects (this problem is treated in chapter 9). At the national level, in addition to possible national funding, there have been many multilateral commitments for energy purposes by the World Bank (over $10 billion in the period 1975-1981), as well as by regional banks (almost $7 billion during the same period) (World Bank, 1983, p. 155). In addition, during the period 1975-1980, the United States, Japan, the Organization of Petroleum Exporting Countries (OPEC), and other industrialized countries committed over $7 billion in concessional grant assistance for energy. Naturally, only a portion of this would be either scheduled for or available to bioenergy projects, though the World Bank in particular has mounted fairly substantial programs for fuelwood development.

None of these figures include the large sums also expended by private, nongovernmental organizations in this field. Indeed, these organizations have often been in the forefront in certain bioenergy projects, like social forestry schemes, for example. Great emphasis has also been placed recently on the development of fuelwood initiatives in AID programs. This aspect of bioenergy use therefore enjoys a relatively favorable funding outlook.

Macroeconomic Factors

Bioenergy programs will often have significant interactions with other sectors of the economy and with such economic problems as balance of payments, terms of trade, and inflation. These effects are complex and often indirect (see chapter 13), but sometimes these factors will have a direct effect on the viability of the bioenergy product itself. Costs for a bioenergy scheme may appear reasonable, and the product may be competitive with other fuels, but if there is no foreign exchange available to pay for necessary capital inputs, the project will be impractical in the real world. An example is the so-called "National Program of Alcohol Fuel Production" in Costa Rica, a diverse set of proposals developed in the years preceding 1982 for replacing gasoline by alcohol from sugarcane. This program could have had a positive long-term balance of payments effect in displacing oil imports. But a key decision factor for energy planners in this case would likely have been the fact that the local value added in such alcohol production schemes was calculated to be less than 50 percent, meaning that the remaining 50 percent involved capital imports that would have to be paid for in the short term with scarce foreign exchange (Celis et al., 1982, p. 56). In the event, a subsequent debt crisis leading to a moratorium on payments spelled the end for the time being of the Costa Rican alcohol option.

In many countries there is a complicated interaction between continual rounds of inflation and periodic lowering of the exchange rate of the local currency against the dollar. This was the case in Brazil during the past decade. Such cycles can introduce a treacherous element into the calculations of costs of bioenergy facilities. If inflation proceeds for some time but the exchange rate is constant, local resources (e.g., one day's labor) will purchase more dollars. The cost of equipment paid for in dollars (or Special Drawing Rights) will therefore appear to drop in terms of local resource equivalents, lowering real project costs. If past experience is any evidence, however, there will be a day of reckoning when the currency is devalued and then the cost in real terms of needed capital imports for any bioenergy facility will make a quantum jump. Whether or not bioenergy projects thus succeed because of inflation and devaluation in competing more successfully with oil depends on the behavior of budgets. If project budgets keep up with inflation or better, or are derived from dollar sources, there may be the possibility of "buying" local value added at cheap rates.

Scale: A Twofold Problem

The question of scale comes up in two different ways. The scale of a specific project in an economic sense will ordinarily be covered in the usual project engineering specifications. Nevertheless, it is useful to examine what different potential returns to scale might be so that the suitability of a selection of projects to particular local situations can be assessed. In addition, a good deal of interest has developed during the past few years in the question of introducing small scale projects for the sake of smallness. This question brings the bioenergy planner into a whole realm of "sociopolitical" considerations. That is, special social benefits such as institution building, participatory democracy, and general quality of life are supposed to be realized by certain innate superiorities of small-scale activities over large-scale projects.

Economies of Scale

Some generalizations usually hold for the question of economies of scale. Other things being equal, large bioenergy conversion facilities, for example, will be cheaper to run—per unit energy output—than the small ones.

This law of economies of scale will follow in developing countries as well as in industrial countries because the causes of the phenomenon are physical and universal. In physical terms, costs tend to follow the surface area, while revenues tend to follow the volume of typical pieces of equipment, and the ratio between surface and volume falls as scale (size) increases. For example, in one application of wood-burning boilers in Costa Rica, boilers producing 5 thousand pounds of steam per hour had capital and operating costs—per unit of energy produced, exclusive of costs of fuel—of about 3.5 times those of a larger boiler producing 500 housand pounds of steam an hour (Meta, 1982, p. 6-2).[3] Results were similar for wood-fired gasification systems feeding similar boilers.[4]

The same tendency can be seen for fuel alcohol. Larger plants can afford more economic distillation columns and heavier rolling mills, which not only enjoy ordinary scale economies of construction, but which also extract more sugar juice from a unit of cane (Netto, 1984a). As far as capital costs are concerned, one review showed that in the region of scaling from about 3 million liters per year to 50 million liters per year there were extensive economies of scale, with economies diminishing rapidly as the 50 million liters a year mark was exceeded (World Bank, 1980a, p. 33).[5]

For production of alcohol from corn in the United States, a range of smaller plant sizes was considered extending from about 40 thousand liters a year to 4 million liters a year (Dobbs et al., 1982, pp. 3-4). Economies of scale were quite significant at the lower end (40 thousand liters), with prices very high at $1.03 per liter, dropping to about $.48 per liter at 670 thousand liters a year, and to $.32 per liter at 1.4 million liters a year. It must also be emphasized that these smaller units produced hydrated alcohol, i.e., did not include the extra step of extracting the final 5 percent water from the alcohol that makes the fuel suitable for mixture with gasoline.

Economies of scale for bioenergy must consider not only the plant costs but also the cost of feedstock delivery. These costs reflect transport costs, and the cost penalty for collecting feedstocks from a distance tend to favor smaller operations centrally located near tree plantations or other crop resources.[6] The effect of this feedstock contribution to the scale factor was not thought to be dominant, however. For the Brazilian plants, the most favorable scale for production was estimated to be in the low-middle range—that is, in the region of 20-40 million liters a year for sugarcane—with a corresponding estimate of 10-20 million liters a year for cassava (manioc) (World Bank, 1980a, p.33)

3. The exact ratio is 3.47, at an 80 percent load factor, with a cost in dollars of $2.84 for the larger boiler. At a load factor of 40 percent the ratio was only 2.5.

4. Another way of evaluating the scale factors is in terms of the price that could be paid for fuelwood to break even for economical operation. At a 60 percent load factor, 11 percent more could be paid for fuelwood for the larger boiler for operation on gas, and 17 percent more for a direct combustion operation (Meta, 1982, p. 6-6).

5. These capacities are converted to yearly rates at 165 working days a year for plants of a capacity ranging from 20 thousand liters a day to 300 thousand liters a day. The following scale factor resulted: for plants producing 80 thousand liters a day, capital costs were 30 percent more per unit output than for plants producing 240,000 liters a day.

6. It is assumed here that the feedstock is renewable, i.e., supplied on a sustained-yield basis.

In the much smaller-scale U.S. case, the model calculations for scale effects showed that the 180 thousand liter per year plant would require 1.5 farms (87 hectares) to provide feedstock, while the 60 thousand liter a year plant would require 5.5 farms (312 hectares).[7]

For biogas, the general results are similar, where a 1250 cubic foot size plant had a benefit to cost ratio of 2.23 (at a discount rate of 15 percent), compared to a ratio of 1.72 for a 200 cubic foot capacity plant and 1.05 for sixty cubic feet (Moulik and Srivastava, 1975, p. 78).

Questions of scale for the cultivation of bioenergy crops are much hazier. Scale factors for such industrial timber operations as integrated sawmill operations have been estimated as requiring a thousand hectares as a minimum-sized harvestable area, while an integrating pulping operation was estimated to require 25 thousand hectares (Evans, 1982, p. 85). There seems to be no obvious minimum economic size for fuelwood operations, however. Indeed, as discussed in chapter 11, small size may in practice be favorable if it allows the use of marginal lands and underutilized labor.

"Sociopolitical" Scale Factors

Over the past decade, energy project scale has become more than just a question of economics. Starting with the writings of the English economist Schumacher in 1973, the idea that "small is beautiful" has gained great attention and has had a strong influence on economic development project design. Perhaps typical of the kind of goals that go into this "sociopolitical" way of viewing scale are those stated in the program of the Center for the Application of Science and Technology in Rural Areas (ASTRA) in Bangalore, India: projects should involve alternative technologies that facilitate low capital investment, employment generation in rural areas, dispersal of mini-production units to the villages, and production of inexpensive goods and services of the mass production variety (Islam et al., 1983, p. 397). In the field of energy policy, this point of view was popularized by Amory Lovins (1977, pp. 97-98) as a "soft path" to the solution of energy problems. Small-scale (or "intermediate-scale") systems were promoted as reducing alienation by fitting in with an existing agricultural system, by being understandable, and by being more immediately relevant to the daily lives of the villagers. Systems such as improved stoves, small biogas plants, improved charcoal kilns, and village fuelwood plantations fit well under this heading. This view has also stressed the ques-

7. These two cases (49 thousand gallons and 175 thousand gallons, respectively, in the original units) would require a wider "service area" for getting rid of the protein byproduct from corn—47 and 168 farms, respectively (Dobbs et al., 1982, p. 7).

tion of equity, with soft technologies involving "more equitably distributed natural energies" to meet perceived human needs directly and comprehensively. Sometimes the wider term "appropriate technologies" is also used to denote systems embodying this approach.

Obviously such a set of policy principles can pose a significant complication in energy policy planning. For one thing, such a prescription is not likely to be politically neutral. As already pointed out, questions of equity are intimately involved in this point of view; in fact, the soft path contains a more or less explicit attempt to redress the balance of economic power (Eckhaus, 1977, pp. 48ff). Perhaps more to the point, the use of smallness of scale for its own sake tends to be tied in with encouraging some types of life-styles over others: specifically, a relatively simple rural existence versus a mass production-consumption urbanized way of life (Eckhaus, 1977, p. 48ff).

Certainly, the promotion of smallness of scale for its own sake has in its favor the fact that there may be sound economic grounds for small, relatively primitive technologies. It is emphasized in chapters 11 and 12 that such technologies can take advantage of underutilized local labor and management skills— and other wasted resources present in subsistence societies— that may be invaluable in new bioenergy projects. For example, it is plausible that household enterprises can make a relatively intensive use of labor, if they work to maximize revenue—less the cost of outside factor inputs bought with cash—regardless of any extra family hours of work needed (Eckhaus, 1977, p. 102).

Bioenergy planners will have to recognize that the "sociopolitical" type of consideration for project scale could be important in project planning—*if* the local population shares the goals of smallness, appropriateness, understandability, etc. In other words, if a local population happens to attach value to the ability to participate in decisions and tasks, to undertake projects on a do-it-yourself basis, and to increase community cooperative activities in general, then this must be taken into account in the planning and in particular in the scale of the project (Dunkerley et al., 1981, pp. 210-211).

On the other hand, local opinion may reflect the disadvantages as well as advantages in widening responsibilities and decentralizing control over energy technologies. Social control may be felt to be actually less efficient. Furthermore, it is not self-evident that the inhabitants of a given community will appreciate the increased responsiblity for social tasks or the necessity for the sometimes prolonged social interaction involved in participatory decisions.[8]

8. It has been pointed out that this kind of judgment is the province of national and local authorities and that outsiders from industrial countries usually have no special qualification to contribute to them (Dunkerley et al., 1981, p. 211).

Fortunately for the planner, small-cale projects undertaken with goals of "participatory democracy" and equity in mind are often compatible with general economic development goals, as long as they truly reflect the needs and motivations of the local community. And in any case community opinions merit the closest consideration in a wide variety of rural biomass energy options.

Conclusion and Planning Procedure: Cost and Financial Outlook and Questions of Scale

Trends of relative costs over the coming decades, as development proceeds in the less industrialized countries, are difficult to predict. Little can be said about long-term costs of labor and capital, for example. There appears to be no reason to think that costs of bioenergy relative to other sources of energy will rise, however, even though large amounts of land—and, in some places, labor—are involved. The uncertainty is most troublesome for land, a key input to bioenergy resources. Candidate cultivation options for areas where land competition can be expected to grow rapidly must be examined critically.

In principle, future prices of petroleum—the major alternative fuel—need to be estimated. In practice this is not feasible. The local outlook for such competitors as imported coal may, however, be clearer—and may affect the viability of a candidate option. Financing, foreign exchange, inflation, and other macroeconomic problem areas may well complicate the analysis. Where the local foreign exchange picture is grave, the estimates of net foreign exchange gains (or losses) on a project can be given highest priority in assessment.

In ordinary economic terms, returns to scale for most technologies in bioenergy follow the usual rule of larger is better. To some extent, these rules have to be bent to recognize that cultivation of biomass may not necessarily be more efficient on a larger scale and that the transportation costs of biomass feedstock to a central conversion facility may make smaller scale plants more desirable than they otherwise might be.

The so-called "sociopolitical" aspects of scale go beyond the normal type of planning in that they set up additional social goals for the development process. The extent to which bioenergy projects should adapt to these goals must depend on local circumstances and on consulting national and local preferences and capabilities.

9

Infrastructure, Marketing, and Distribution

Introduction: The Outside World

The bioenergy project does not exist in an economic cocoon. The project itself will require support from the network of outside institutions and social entities usually labeled "infrastructure." And once the project produces a feedstock or a fuel, the product must be marketed and distributed. Often these areas are outside the control of the project planner; however, they are always of concern to bioenergy development scenarios.

Infrastructure: The Anomaly of Project Incompleteness

In bioenergy planning as in other forms of economic development, the question of infrastructure involves a logical anomaly. "Infrastructure" generally refers to elements in the economy—like roads and schools—that may be necessary to the success of some form of economic development project, but are not normally included as an integral part of a project design. In a sense, therefore, infrastructure means something that could have been part of the project—and in too many cases *should* have been—but was not. The reason for this omission could be poor planning. On the other hand, much more often the infrastructure element—like a railroad line—is so expensive that the project must depend on someone else to supply it.

Bioenergy projects can require just as much infrastructural support as any other type of economic activity if not more. The term "infrastructure" can include anything, even such social infrastructure as schools and hospitals. The most direct effects on bioenergy projects, however, have to do with categories like transportation, maintenance, credit, and training.

First, we examine "quasi-infrastructure," or infrastructure that is often included as part of a project; then we look at the problems and opportunities involved in other types of backup facilities needed to insure the success of bioenergy projects.

Quasi-Infrastructure: Subsidiary Project Requirements

Elements that are called infrastructure for some projects are explicitly included as elements of the project in other cases. This is especially true for many small-scale efforts in developing countries, where projects designed to stimulate development in rural subsistence economies cannot count on much secondary support in transport, credit, or other societal elements essential to project success. Therefore, the bioenergy planner will often include such key factors in his project. For example, in a gasification scheme in the Philippines, the Farm Systems Development Corporation (FSDC), the government "umbrella" organization, was to provide a number of technical services to farmers participating in the scheme: operating and maintenance help for the gasification equipment, instruction in tree production techniques, managerial training for woodlot operators, and a financial system for repayment of loans (AID, 1982, Annex II, p. 7). In addition, the FSDC undertook the essential planning role of synchronizing fuel production with irrigation.

Roads are one element that may have to be included as an integral part of the project. This is particularly true for forestry projects: good roads may not exist, the geographical area covered may be rather large, and difficult topography will make the delivery of the harvested wood costly. This is an old story to timber companies operating in the developing countries, where road design forms an important part of most projects (FAO, 1976a, pp. 13-25). The situation is more complex (expensive) for fuelwood projects, however. Primitive roads may be adequate for harvesting natural forests. But for the longer term needs of plantation forestry, it is recommended that the feeder roads running from the secondary road system be designed and built as permanent roads capable of carrying the expected gross haul vehicle weights over extended periods (FAO, 1976a, p. 15).

Infrastructure Proper

In many cases, infrastructure will have to be already present, or the project will be impractical. In even more cases, pilot or demonstration projects will include transportation and training as part of the project itself. But replication of the project in other locations within the country may be impossible unless the requisite social support structure already exists.

Some infrastructural elements constitute absolute requirements for the project. Other pieces of infrastructure are not indispensable, but their presence will improve the economic feasibility of the project.

Transportation. Although many areas in the developing world will have transportation networks—either road, rail, or water—that can provide an adequate base for bioenergy projects, the situation of concern to the planner is those areas where transportation is carried on by exceedingly primitive means. Where walking, bicycles, or animal transport are the principal methods of moving goods and services, transportation costs can be a severely limiting constraint. And it is an expensive constraint, no matter how it is considered. First class roads can easily cost several hundred thousand dollars a kilometer, and even much more modestly constructed rural roads cost several thousand dollars per kilometer (Ramsay and Shue, 1981, p. 241).[1] Cheaper road design is no simple solution. Although narrower and more poorly surfaced roads would cost less and existing rural tracks could be improved quite cheaply, poor roads incur heavy financial penalties in operating expenses—mainly truck operating costs and driver salaries.[2]

These problems are, of course, the same that have plagued the development of tropical agriculture. Indeed, the existence of rural road networks obviously play a familiar facilitating role for bioenergy feedstock production systems—like tree plantations and sugarcane fields. And in practice, bioenergy projects may sometimes have to piggyback on agricultural programs or programs for "integrated rural development" in order to justify any needed new transportation facilities. On the other hand, large commercial bioenergy ventures like fuel alcohol distilleries or region-wide direct combustion electricity schemes may have enough local impact to stimulate the development of transport infrastructure throughout the region.

Infrastructure for distribution and marketing is an important problem covered in the next section.

Credit. There are several levels to the credit problem. We have already briefly considered (in chapter 8) the problem of overall project financing. The provision of credit at the level of the energy producers or users is exceedingly important, however, not only for particular bioenergy schemes like household biogas digesters, but also for the replication of demonstration projects of producer gas for irrigation pumping, for example. For larger lenders like commercial-scale alcohol projects, for example, such funds may be directly obtainable from international donors or large banks. For smaller projects carried out by farmers or other smallholders or by community-level organizations, funds may also sometimes be obtainable through donors or national

1. A typical 5.5 meter laterite-surfaced feeder road in Mali (designed for 4 thousand tons or more per year of travel) cost about $2,450 per kilometer in 1970.

2. The average transport cost on feeder roads in Mali was reported as 60 percent more than on paved roads, while rural tracks were assigned almost a 400 percent higher cost of operation than paved roads.

development or commercial banks. In particular, over the past few decades commercial banks in India have increasingly become a primary source of rural credit. Specifically, the decentralized offices of such institutions as the Syndicate Bank, with its program of "pigmy deposits," have helped to facilitate rural loans and encourage rural savings (Ramsay and Shue, 1981, p. 236).

On the other hand, in many places credit may be difficult to come by on a small scale. In 1971 in Kenya, some 88 percent of all institutional agricultural loans went to large-scale farmers, leaving little credit to be distributed among small farmers (Ramsay and Shue, 1981, p. 237). In some places, this problem can be circumvented through the formation of farm cooperatives and other groups; one group in Kenya organized a "circular loan" system among its members by collecting dues and distributing them to different members in turn (Lele, 1974, p. 83). Sometimes special credit arrangements are available from specific bioenergy agencies such as the Khadi and Village Industries Commission in India whose program has involved mostly funding to individual farmers on a household scale (Ramsay and Shue, 1981, p. 239). And traditionally, private money lenders have been a primary source of credit in many rural areas.

This credit problem may be central in the viability of a bioenergy program, and some of the lessons learned in agricultural credit can be usefully studied by the bioenergy planner. In particular, one can note that the high interest rates typical of money lenders may represent real risk premiums involved in loans in a marginal cash or a purely subsistence economy—i.e., default rates may be high, so interest charges have to be high to cover losses. Successful credit programs may therefore have to involve high interest rates to be financially viable to the lender (Ramsay and Shue, 1981, p. 238).

Maintenance and Repair. Lack of maintenace and repair facilities have very likely contributed to the problem voiced in the complaint that "many small-scale energy projects in LDCs . . . have been abandoned" (Howe, 1977, p. 35). Inadequate maintenance has been the cause of "disasters" in agricultural tractor schemes, for example (Singh, 1977, pp. 19, 31). And the necessity for teaching technical know-how in businesses like auto repair shops and other trades relevant to the energy field has been recognized in AID program planning (Lele, 1974, p. 165).

Advocates of "appropriate technology" for developing areas have stressed that systems that are sufficiently simple can be repaired and maintained locally—at least in principle. The practice may be another story, however. Therefore, it is interesting to speculate that systems requiring *constant* maintenance could be considered best suited for environments lacking a reliable maintenance infrastructure (Hirschman, 1958, pp. 138-152). If maintenance must be carried out continuously (as for airlines), the contention is that repair and maintenance become a normal part of operations and cannot be neglected—as they can for systems requiring less frequent repair (e.g., railroads). Programs promoting improved charcoal kilns could display this effect: primitive kilns require only a very simple kind of maintenance but the maintenance is essentially continuous, while more elaborate kilns could suffer from on-again-off-again attention.

For biomass cultivation and feedstock delivery schemes, the planner could be faced with problems such as possible malfunctions in mechanized harvesting equipment as well as general maintenance problems and the transport infrastructure problem of keeping necessary roads in repair.

Training. The subject of training receives a good deal of recognition as essential for project continuity and replication. Certainly, in community fuelwood schemes the important role of forestry department staff in training villagers to plant, tend, and harvest trees is widely recognized. In particular one study of certain villages in Tanzania found evidence of the pivotal role played by the local ranger (Bwana Miti) in training the villagers through extension services (Skutsch, 1983, p. 34). The majority of the foresters observed in a gum arabic agroforestry project in the Sudan worked effectively to train local farmers to make the program a success (Hammer, 1982, p. 51).

The subject of a training infrastructure is often intimately connected with the question of maintenance. The Chinese biogas program, for example, has been characterized by a concerted policy to have maintenance technicians train other technicians through an apprentice program (Van Buren, 1979, p. 125).

Nevertheless, for bioenergy as a whole the provision of special "energy extension" services has remained up to now largely a dream (Howe, 1977, pp. 35-36). The bioenergy planner will, therefore, in general be forced to deal with very simple programs or with programs that are large enough to warrant hiring professional operators and maintenance staff, or will have to set up explicit training mechanisms. Sometimes these training schemes will have to be rather extensive in scope. A recent gasification scheme for irrigated farming in the Philippines involves training the members of the farmer associations in operating and maintaining the irrigation pumps and teaching the tree-growing laborers about tree cultivation and wood chipping (AID, 1982, Annex II, p. 7).

Marketing and Distribution: Planning By Indirection

Getting new biofuels distributed and into efficient markets for delivery to the energy user is not only one of the key links in the whole bioenergy system, it is also one of its most treacherous problem areas. The bioenergy planner or producer often has little or no control over markets and distributional infrastructure. Therefore, the problem becomes an exercise in guessing what the behavior of others—truckers, brokers, retailers—will be. As the discussion above has suggested, even where a project manager is directly responsible for at least part of the infrastructure (transport for example), he will tend to make his decision on what is best for production—and the best transport system for production may be not the best for distribution.

With markets, fragmentation of enterprises—with a large number of different dealers—is often the rule, making it difficult to coordinate improvements in marketing and distribution. Even in centrally-planned economies, both production planning and market planning may not be sufficiently "centralized"—individual state enterprises may be effectively resistant to central control.

The problem, then, becomes more one of understanding how new (and old) biofuels may fare in marketing and distribution systems that may or may not be subject to outside interventions like new market subsidies or highway improvement projects.

Existing Patterns of Bioenergy Distribution and Marketing

In many parts of the world there is no settled bioenergy "market" in the usual sense; fuelwood or wastes are gathered by individual households from forests or fields that are either unclaimed or are treated for gathering purposes as community property. The "costs" then have to be measured in person-days needed to collect fuelwood, in related drudgery and fatigue, in the temptation to cook fewer meals with a generally negative effect on nutrition and in increased use of agricultural wastes as fuel rather than as soil amendments or fertilizers.[3] The kind of noncash transactions by which much of present bioenergy is gathered or exchanged is therefore an important fact of

3. This type of activity has been an intensely studied topic, especially in its connection with the desertification of areas in the Sahel and the localized gradual disappearances of fuelwood resources around settlements—whether due to fuelwood-gathering pressures on the forests or to the clearing of forests for agriculture, or both. For example, it has been reported that 250-300 person days a year are needed to supply wood for a family of five in parts of central Tanzania, and in recently deforested areas of Nepal, that 5-19 person days per month are spent in collecting wood (Spears, 1978, p. 6). Wood collection has reported to take up one out of every four days in some parts of Benin, with walking distances of up to 15-20 kilometers from the village being necessary to gather supplies (Woron and Tran, 1982, p. 8). (A "market equivalent" calculation for this case held that a bundle of wood to supply the cooking needs of three days cost a day's wages.)

life that must be dealt with in planning a more modern bioenergy economy.[4]

Real-life bioenergy markets, however, are sometimes considerably more complicated than this simple picture. In one province in the Philippines (Ilocos Norte), a detailed survey showed that many fuelwood users grew their wood supply while a much smaller number collected "free" wood (not from their own land). Many users either bought cut wood or paid a landowner for the privilege of collecting wood on his property.[5] Therefore, this one market involved no less than four means of acquiring fuel. The survey also found out that the majority of fuelwood *sellers* got the wood from the property of their landlords.[6]

Traditional biofuel markets are not all rural; cities also use significant amounts of fuelwood and charcoal, and urban biofuel markets can be more complex than their rural counterpart. We have already mentioned (in chapter 10) an in-depth study of the supply of fuelwood and charcoal to the city of Hyderabad in India. The study showed several idiosyncrasies in the modes of getting supplies distributed to a city of 2.1 million people. Fuelwood was transported into the city mostly by truck from an average of 75 kilometers away.[7] Eighty percent of fuelwood came from private lands, usually farms, but about 20 percent came from government forests. Then a complicated market system took over. The fuelwood sellers, who contracted with the government or other private owners for the wood and transported it to the city, brought it to a large marketing area. There, sales were not made directly to retailers and to individual users[8] but through middlemen, who arranged the sales between the fuelwood truckers and the customers.[9]

4. The inefficiencies of "nonmarket" societies (barter and gathering economies) feature prominently in the sometimes anguished controversies about deforestation and rural living standards. For the bioenergy planner, deforestation is strictly speaking a supply, and not a marketing problem in itself. This is not to take sides in the sometimes acrimonious debate about deforestation: the extent thereof, the reasons for, the possible cures, and priorities vis a' vis other development problems. For a reasoned view of this situation see, for example, Allen and Barnes, 1981.

5. Hyman, (1982a, p. 7) reports that 39 percent of the respondents grew 100 percent of their own wood supply, about 6 percent collected free wood, 30 percent collected wood at a contracted price, while 17 percent bought all of their wood already cut. There were also other smaller percentages representing persons who bought some wood but collected or grew or contracted for other portions of their needs—and all permutations of these possibilities.

6. Eighty-one percent, according to Hyman (1982a, pp. 12,14). Two-thirds of these fuelwood sellers reportedly spent less than 12 days a year acquiring the wood. In contrast to many studies emphasizing the women's role in gathering fuelwood for households, for these fuelwood sellers only 35 percent of the households had involved the collecting by women (and children, 13 percent) while males collected in 76 percent of them.

7. The furthest source of fuelwood was 280 kilometers away. Ninety-five percent was brought in by truck, the rest by bullock cart. Statistics on the movement of bullock carts may be somewhat underestimated because of counting problems (Alam et al., 1982, p. 10).

8. Large individual users tended to buy through the middlemen, not through retailers. Large individual uses included cremations, wood being the obligatory fuel.

9. The rationale for the middlemen is not immediately obvious: they did not seem to be necessary for the charcoal market, for example (see below) (Alam et al., 1982, p. 20).

The structure of the charcoal market in Hyderabad was entirely different. First, the production of charcoal was formally regulated, the state government requiring a charcoal-making license—a widespread practice in the developing world. Second, the charcoal came mostly from larger distances, often from outside the state, and most of it was brought in by railroad.[10] Most of the customers, retailers, and individuals bought it directly at the railroad depot.

In Mogadishu, Somalia, fuelwood harvesting and distribution to firms and institutions is handled by the national cooperative "Galool" (Smale et al., 1984, p. 6). Members of the cooperative hire and supervise the harvest laborers and transport the fuelwood in trucks to town. The cooperative also produces charcoal and sells it wholesale to local shops. Fuelwood for households, however, is supplied by individual private producers and traders.

Khartoum, located in the middle of a semi-arid region of the Sudan, supports a large market in fuelwood and charcoal (Sudan, 1982, p. 2). Although the detailed mechanics of the market have not been studied, some gross effects have been noted, such as the impact of higher transport costs on local biofuel prices as a result of local deforestation.[11] In fact, recent increases in the market cost of fuelwood and charcoal were so high that charcoal had actually become more expensive than kerosene—usually considered a superior fuel—for cooking in Khartoum.[12]

This brief overview of some present markets should give a sense of some of the problems involved in planning changes in distribution and marketing patterns for biofuels in the future. Improved distribution networks might make sense for new forms of biofuels or for greatly expanded use of older forms, but it should be recognized that present marketing institutions can represent a formidable amount of structural inertia in the economy.

Outlook For Bioenergy Market Patterns

We have already seen in chapters 2 and 6 how biofuels can substitute for fossil fuels in numerous critical end uses in developing countries. And of course new and improved forms of biofuels can often substitute for the old and reliable—but sometimes inefficient—use of wood and charcoal both in households and in rural commerce and industry. In

10. The average distance was 217 kilometers, with 70 percent brought in by rail (Alam et al., 1982, p. 30).

11. In another area (Costa Rica), minimum costs for transport of wood were estimated at 0.4 cent per gigajoule-kilometer and 0.8 cent per gigajoule-kilometer for charcoal (Meta, 1980, p. 7). One gigajoule (109 joules) is 0.94 MMBtu, or about the amount of energy in seven gallons of gasoline.

12. Even fuelwood in Khartoum cost somewhat more than kerosene per unit of energy supplied (Sudan, 1982, p. 3).

order to get some idea about whether new distribution systems and markets are needed to handle increases or changes in the use of biofuels, the bioenergy planner needs the most informed estimates available on the future market volume of traditional fuels. Development can change patterns. Because charcoal has always been made *in situ* in the Mbere Division of Embu District, Kenya, new roads provided a radical stimulus to charcoal production and marketing (Brokensha and Riley, 1980, p. 23). Furthermore, changing patterns in biofuels production mean changing patterns both in agricultural activities affected by bioenergy land use and in industries and households dependent on biofuels.

The projection game is not easy. The present pattern of use of wood and wastes for fuels, especially in rural households, is known in a quantitative way only from a scattering of case studies. In order to make general projections of future rural household use, one ordinarily guesses in a simple-minded way that the per capita use of "traditional fuels" (wood and waste) will remain the same, so as rural populations get larger it is assumed that rural fuelwood and wastes consumption will increase correspondingly. As we noted before, this kind of projection has obvious shortcomings.[13]

The data situation is only marginally better for commercial and industrial use. One overview has suggested that the trend has been for the use of traditional fuels in industry to increase in absolute terms—and also for industry to take over an ever-larger share of total use of traditional fuels (Dunkerley et al., 1981, pp. 56-57).[14] But at the same time, there is a another overall trend—quite strong in several countries— for new additions to the industrial sector to use fossil fuels instead of traditional fuels, meaning that the percentage use of traditional fuels in industry should decrease. Therefore there could be no net increase in industrial traditional fuel demand. On the other hand, if new initiatives to increase the industrial use of new (and old) biomass fuels were to succeed in bucking this last trend, it may mean very large demands on the biomass resource. For instance, total pig iron production in the Third World in the late 1970s was about 80 million tons a year. If that tonnage had all been produced by charcoal, that would have meant a 50 percent increase in the demand for industrial wood (Smil, 1979, p. 15). Such increases, *if they occurred*, could

13. One shortcoming is the neglect of rural migration to the cities, which could, depending on the precise demographics, change the balance of fuel use: even though urban dwellers use a good deal of traditional energy, they tend to use less per capita than rural inhabitants. In the second place, especially in places like Latin America, more modern options like electricity are tending to replace fuelwood in rural cooking.

14. In fact, it is rather striking that in some countries, such as Thailand and Argentina (see Dunkerley et al., 1981, p. 56), most traditional fuels are (reportedly) used in industry, not in households.

then put significant demands on present supply facilities and distribution networks for bioenergy fuels.

Finally, if bioenergy becomes more widely used, the marketing and distributing of by-products have to be considered. By-products could be considered either a problem or an opportunity. Increasingly, bioenergy schemes that are marginal from the point of view of energy use itself can show very favorable economics as part of a larger economic development project, utilizing by-products in creative ways. In Fiji, for example, a proposed biogas system was analyzed as uneconomic, until it was included as a part of an integrated farming system; local farm product markets then become of keen interest to the energy planner.[15] Schemes for improving the economics of other bioenergy fuels have also looked to by-products for help. New processes have been developed for modifying the cane milling process, for example, so that the residue is split cane that can be used for building materials, rather than crushed bagasse.[16] Other types of biofuel programs have developed other types of by-products, some of them rather exotic, such as kudzu bark, used worldwide as a wallpaper material.[17]

New Challenges in Distribution and Marketing

Any new distribution and marketing systems introduced for the benefit of a new, expanded biofuels economy must obviously work to some extent within existing systems. But biofuels may stimulate, or benefit from changes in, national distribution and marketing patterns.

Distribution. Distribution systems will often cause no problem for a new biofuels economy. If there is a problem in distribution connected with biofuels, it is quite likely a problem we have already looked at in the discussion in this chapter on infrastructure for production. For example, a new village woodlot may require paths or roads to reach the lot or to plant the trees, but the same road can also be used to bring the wood back for use in the village.[18] In the same way, in local tree-growing schemes sponsored by electric utilities, wood has to be moved to the power plant, but existing roads into rural population

---

15. As reported by Chan (1981, p. 127), the effluent from the biogas plant was fed into a plastic-lined algae basin; in turn, effluent from that basin was sent into a deep pond where fish and duck were raised; the effluent from the second pond was then used to irrigate and fertilize crops (see Chapter 12).

16. As remarked in chapter 7, stillage—leftovers from the distillation process—is a pollutant when dumped in rivers but a fertilizer when spread on fields. Such chemicals as furfural $C_5H_4O_2$ and fusel oil (mainly amyl alcohol, $C_5H_{11}OH$) can be valuable byproducts from fermentation and distillation— especially if they can be successfully utilized in the local manufacture of plastics, for example (see, for example, Ramsay, 1979, p. 60).

17. Also in Korea, mushrooms from both oak and pine forests can be either grown from spores injected into logs, or will occur naturally: they are a valuable export crop, especially to Japan (FAO, 1982a, pp. 99-100).

18. The same thing happens when no distribution is needed, when wood is delivered to a gasifier to run a water pump, for example. In that case, technically speaking, the biomass conversion mechanism (the gasifier) is already at the "point of end use" (the pump).

centers will presumably be utilized. Furthermore, even if such roads are not available, the building of the road will normally be included as part of the project design and would not show up as a separate "distribution problem."

Other, more exotic bioenergy systems may indeed encounter some distribution problems. Community biogas systems, for example, can have problems of gas distribution, or perhaps of gas distribution and pricing—who gets how much at what cost.[19]

One would expect, on the whole, that either the market system itself, or fairly obvious government actions would provide adequate roads and other means of distribution for bioenergy delivery needs. Certainly, an entrepreneurial market in charcoal in the Sudan is carried on with apparent efficiency by long-distance trucking. But it must be cautioned that, as some evidence from the Brazilian fuel alcohol program attests, distribution systems do not take care of themselves. It has been reported, for example, that access roads to distilleries in Brazil are mostly unpaved and are often in a precarious state, especially in the wet season.[20]

Markets. Markets always present complex problems. The planner may hope that adequate markets will develop automatically, but in this complicated economic world one can never be sure. Therefore, investigation is necessary before investment. It is probably a safe guess, however, that more renewable energy projects have come to grief through omitting a thorough investigation of potential markets than through any other cause. Bioenergy planners therefore neglect the marketplace at their own risk.

Forestry plantations probably present some of the worst problems. To be sure, some plantations are village-type woodlots —which supply a guaranteed market in one specific community. But even a community woodlot can have leftover wood to sell: in that case they—in company with government or commercial plantation schemes—may find that lack of an assured market can be the downfall of a promising forestry project (Evans, 1982, p. 94). The long-range aspect of tree growing makes the problem even more difficult, because the marketing facts of life can change radically during the life of the project.[21]

---

19. Problems have been anticipated in such systems as to whether gas should be distributed freely to members of the community or whether gas metering—which is expensive—is necessary. Similarly, it may be difficult to devise an equitable distribution scheme for the sludge (fertilizer) residue (Moulik and Srivastava, 1975, p. 52).

20. There are also projections that storage tank capacities for ethanol will be insufficient for the amount of fuel alcohol projected for production in the 1980s (MT-GEIPOT, 1979, p. 102).

21. In a nonenergy context in Swaziland, a forestry company began planting in the mid-1960s to produce eucalyptus mining timber. But mining timber had become a glut on the market by the time tree harvesting began 10 years later, and the timber ended up being sold mostly for particle-board manufacture.

Other market problems can be caused by "stickiness" in the movement of resources into the biofuels area. A recent study in Costa Rica analyzed one bioenergy market there as being characterized by a surfeit of sellers with very few buyers. The reason offered was that knowledge and capital are needed in order to be able to introduce wood waste biofuels as a replacement for oil (Meta, 1982, p. 7-7). The same study, however, seemed to be optimistic about the ability of free enterprise to correct those market problems in time (Meta, 1982, pp. 7-2, 7-4). A great deal of this kind of analysis must be educated guesswork because the planner cannot usually predict the responses of present users of fuels to possible price changes either of fossil fuels or of renewable fuels like biofuels. Consider for example the response of users of kerosene to rises in the price of petroleum products. If higher prices for kerosene discourage its use, it may be much easier to supply the suburbs of cities with such alternatives as biogas.

Some apparently ingenious attempts have been made to insulate bioenergy sellers and buyers from the vagaries of the open market or to protect them from over-dependence on a single outlet, however. For example, under the PICOP scheme in the Philippines (see chapter 12), smallholders growing trees for PICOP are allowed to sell the wood on the open market at higher prices than specified in their contract with PICOP—but the PICOP organization reserves the right to match any such outside offer (Hyman, 1982, p. 4).

Electricity markets are always difficult to deal with. The principle of letting cogenerators and other small suppliers of electricity sell excess electricity to the utility has been fairly well established. The *practice*, however, has been difficult. Getting a good price for this electricity has meant the difference between success and failure for firms like Hawaiian sugar mills that have set up cogeneration schemes—using sugar "trash" and bagasse fuels—to generate their own electricity and to sell the excess to the utilities.[22]

Perhaps one of the most highly publicized manipulation of the markets for renewable fuels has been as part of the Brazilian alcohol program, PROALCOOL—as mentioned in chapter 10. From the retail market side, we may note that the National Petroleum Council has given alcohol an exemption from the federal fuel tax and fixed a price parity between gasoline and alcohol (Dias, 1979, p. 22). Furthermore, the auto registration tax was reduced for alcohol cars and financing was provided over longer terms than for gasoline vehicles (Fagundes Netto, 1980, section 4). Such radical subsidization schemes evidently

22. Part of the problem in a recent series of projects (Castberg et al., 1981, p. 82) was to get the utilities on the big island of Hawaii to buy electricity on a firm basis, just as the utilities had traditionally bought firm (relatively high-priced) power from sugar factories on Kauai and Maui.

work; owing to all these policies, 60 percent of all the cars sold during the last half of 1980 were (pure) alcohol-fueled vehicles (Goodrich, 1982, p. 5).[23]

Conclusion and Planning Procedure: Infrastructure, Marketing, and Distribution

The importance of infrastructure, markets, and distribution for the candidate bioenergy project will depend on the type of project. A scheme for producing alcohol from fermentation of sugarcane and using it for blending in gasoline, for example, will ordinarily have to encompass in the project design itself a great deal about such matters as maintenance and transportation. At the other extreme, forestry projects that involve distribution of seedlings to private individuals may have almost no concern with infrastructure, yet the question of the availability of infrastructure (credit, for example) may be critical in whether the project is a success or not. For example, it is conceivable that projects for the gasifaction of crop wastes could fail to forsee the extent of future needs for repair.

A possible pitfall lies in a conflict between encouraging individual initiative and making sure that the project goals are properly implemented. Schemes involving farmers or small entrepreneurs, for example, may have to ensure that local credit facilities are available and that local transportation facilities will be likely to respond to increased needs for marketing.

Each type of biofuel, old and new, in each kind of economic and social environment may present its own special distribution and marketing problems. In some cases, the free market may actually work, and there will be no problem. In other cases, without government intervention, the biofuel project will inevitably fail. Therefore, intervention in building distribution networks, or vehicles, or in guaranteeing markets, may be necessary to promote the product of a given biofuel option; however, costly mistakes can be made easily, and bioenergy planners must give serious consideration to the appropriateness of specific government interventions in distribution and marketing.

23. Of course, the problem with such effective government subsidies may be whether the subsidies are justified in the first place (see, for example, Islam and Ramsay, 1982, p. 20).

10

Large-Scale Commercial and Nationalized Enterprises

Introduction: Products and By-products

Bioenergy has traditionally been a small-scale activity. There have been exceptions, but wood, animal and crop wastes, and other "traditional fuels" have usually been supplied from small farms, often as incidental by-products of agriculture and tree growing for other purposes, or from small-scale harvesting in unowned or government-owned forests. But in expanding and improving the use of bioenergy in developing countries, larger scale commercial and government-operated enterprises need to be considered. Newer types of energy forms such as alcohol fuels and biogas in particular might benefit from the technical economies of scale and improved management skills possible in larger enterprises. Furthermore, the bioenergy planner must take into account the use of residues from existing large commercial enterprises—such as sawmill wastes—and other cases where biomass energy could be a by-product of other operations for food, fiber, or other purposes.

Large-scale Commercial (Private and Governmental) Bioenergy Production

Existing and planned bioenergy operations on a commercial scale range from the large-scale exploitation of natural forests of fuelwood and charcoal to biofueled steel factories and large alcohol distilleries. These operations are often run by government entities of one kind or another. This section treats these enterprises that are operated as a *business*. Other government-sponsored bioenergy schemes are run by the government as a *developer* or agent of change; they are treated in chapter 12.

Fuelwood and Charcoal From Government-Managed Forests

A study of the fuelwood and charcoal system supplying the city of Hyderabad in India has described some instructive features of a bioenergy system that is "mixed" both in the sense of combining government and private aspects and in displaying differences in scale within the same industry. How inevitable the complications like those in the Hyderabad case are is not clear, but the knowledge that such complexities exist and must be dealt with can be useful to the bioenergy planner in trying to adapt the design of large-scale wood and charcoal schemes to fit national needs.

The government of the state of Andhra Pradesh, in which Hyderabad lies, is both the owner and administrator of a large part of the forests, a regulator of the private forests, and furthermore is in the fuelwood cutting and selling business itself. Trees may not be cut, except by permission of the government, in the so-called "protected" forests of the state.[1]

Two methods for extracting and disposing of wood from government forests have been followed. Under the first method, the standing trees are sold at public auction. The winning bidder pays and then is required to cut and transport the trees away within a specific period of time. This method of organization of production has experienced many problems and has largely been abandoned within recent years.[2] The second method, now more generally followed, is that the government department fells the trees, cuts them into logs, grades the wood by type and quality and then transports the better quality wood (for construction purposes) to a central depot for auction. The lower quality wood is auctioned at the site for fuelwood (Alam et al., 1982, pp. 11-17). The government thus becomes the prime producer of the wood from government land.

A great deal of the fuelwood delivered to Hyderabad is grown on private land. The government regulates the rights to extract on private land and issues transport permits for the harvested wood.[3] The local fuelwood industry is therefore a mixture of contractors and others getting wood from private forests and contractors either clearing trees from government lands under license or buying trees for fuelwood use that have already been harvested by the government departments.

For charcoal, the situation is somewhat different. Although most of the fuelwood consumed in Hyderabad is produced within Andhra

1. The State Forests are classified into three categories. "Unclassified" forests are open to the public. "Protected" forests are open to individuals for grazing purposes under licenses granted for a fee by the government, and trees may only be cut by government permission. "Reserved" forests are classed as inaccessible to use by private individuals (Alam et al., 1982, pp. 3-17).

2. Complaints against this method include that of collusion among contractors to vitiate the bidding process; sometimes more wood was cut than agreed; complaints were made about inadequate wages; and cutting schedules were not followed, thereby interfering with government reafforestation planning (Alam et al., 1982, p. 20).

3. The law requires a good deal of red tape. The owner of the land needs to get a certificate from the revenue department with detail about ownership and rights of the trees on the land. Extraction can begin only after an assessment of the quantity of wood on the land and after completing other procedures laid down by the Board of Revenue (Alam et al., 1982, pp. 29–31).

Pradesh State, 85 percent of the charcoal is produced outside it; 17 firms supply the market (Alam et al., 1982, pp. viii-v). As in many developing areas, a license from the State Forest Department is necessary in order to manufacture charcoal (Alam et al., 1982, pp. 3-10).

Large-scale Tree Plantations

"The biomass farm" has long been a dream of energy planners (De Beijer, 1979, pp. 27-29). In theory, short rotation harvesting on a sustained basis, using, for example, some of the species considered in chapter 4, should be able to produce a relatively inexpensive source of energy. In practice, however, the idea has been less than successful because the returns to investment have not been sufficiently large at present energy prices. In particular, landowners have typically been resistant to forestry plantations for fuelwood purposes because of the long times involved between planting and harvest and therefore the remoteness of the benefits. Therefore, in most areas neither farmers nor large commercial enterprises have shown much interest in forestry for "nonindustrial" (nonfiber) purposes such as fuelwood (Evans, 1982, p. 47).

This situation may or may not be changing. Tree plantations of sizes up to 80 hectares have been reported in the Indian state of Gujarat; these enterprises reportedly produce a number of products such as dates and other fruit and poles for construction, as well as fuelwood (Java, 1984a). Their viability undoubtedly has been strengthened by the aggressive state program of silviculture (Noronha, 1980, pp. 6,12). Indeed, some have suggested that the effects of state forestry subsidies have distorted production away from food crops to production of wood for construction (Deutsch, 1984). Other tree-growing firms on a significant scale have been reported near Addis Ababa, in Ethiopia (Skutsch, 1984). Furthermore, at a smaller scale, there is a large amount of tree-growing for fuelwood and other purposes by individual farmers and other smallholders; these cases are treated in chapter 12.

One outstanding exception to the lack of interest in large-scale fuelwood schemes has been the growth of bioenergy forest plantations as part of a vertically integrated industrial system, in which the biofuel is fed directly into large enterprises. In particular, the state of Minas Gerais is both the steelmaking center of Brazil and the location of more than one-third of the total reafforestation occurring in Brazil between 1967-1978 (Sedjo, 1979, p. 304). The connection is that charcoal is used in the local steel industry instead of coking coal, which Brazil would have to import. This industry supplies the major part of Brazilian pig iron and steel market. Eight large private and two parastatal integrated steel firms produced $1 million tonnes of pig iron in 1976 (Meyers and Jennings, 1979, pp. 48, 234). These firms produce charcoal from their own eucalyptus plantations—the firm Belgo Mineira employs 8300 people in reforestation and charcoal-making—with supplemental purchases from independent charcoal-makers. From the point of view of the bioenergy planner, it should noted that the growth of this industrial pattern has been stimulated within the last two decades by strong government incentives: industrial companies can substitute investments in "approved projects"—especially reafforestation projects—in lieu of payment of up to 50 percent of their tax liabilities (Sedjo, 1979, p. 2). However, 2 million tonnes of pig iron were also produced by 60 smaller (mostly family) firms in 1976. Most of these purchased charcoal from small charcoal burners using scrub forest wood or from reforestation specialists, where tax breaks should have had less impact (Meyers and Jennings, 1979, p. 51).

Charcoal use in iron smelting is of course old and was only discontinued relatively recently in Sweden in the face of cheap imported coal. One plant exists in Australia, operating off wood culled from logged forests, producing 60 thousand tons per year of pig iron; a 275 thousand tons per year plant in Argentina uses both native brush and plantation-grown eucalyptus (Meyers and Jennings, 1979, pp. 183, 199, 209).

Electricity From Wood

One of the most striking examples of a large-scale bioenergy enterprise is the recent "dendrothermal" (electricity from wood) program in the Philippines. The goal of the dendrothermal program is to produce a network of 70 power plants fired entirely by wood chips to produce 200 megawatts of power by the year 1987 (AID, 1982, Annex II, pp. 1-4). By the third year of the program, some 5 thousand hectares had been planted in *Leucaena leucocephala* (Denton, 1983, p. 222).

The organization of the program combines a national agency, local electric cooperatives, and small rural associations. The National Energy Agency (NEA) organizes Tree Farmers' Associations (TFA) of 10-15 members, which are to grow, harvest, and supply wood at power plants operated by collection points to Rural Electric Cooperatives. The NEA supplies the TFAs with 100 hectares of land on long-term inexpensive leases and will see that roads and other necessary infrastructure are developed by appropriate government agencies. The TFAs are financed through crop loans from the NEA that are passed through the cooperatives. The TFAs will receive a guaranteed price for the wood from the utilities. The associations will also share the management of the power plant operation, along with the NEA and the Rural Electric Cooperatives.

One key to the successful organization of the program thus far has apparently been initiative at the local utility level (Denton, 1983, pp. 49-66). The interaction of the utility (cooperative) and the NEA with the local associations (see chapter 12) is also a crucial nexus, combining questions of technical feasibility with social customs and revenue management. For example, experiments have been carried out with corn growing in among the *ipil-ipil* (leucaena) seedlings to supplement "tide-over" loans; there has also been some income available from selling leucaena seeds and tree trimming residues (Denton, 1983, pp. 60-63).

Fuel Alcohol Schemes

The largest and the most successful example of commercial bioenergy production in the world is the Brazilian alcohol program—PROALCOOL.[4]

A number of other countries have made some efforts in this field, often under the direct influence of the Brazilian experience. Zimbabwe has a 40 million liter per year plant that supplies enough alcohol for a 15 percent gasohol blend to replace gasoline in the nation's motor fuel supply (World Bank, 1982d, p. 24). In the Philippines, 14 new distilleries are planned under their "Alcogas" program based on existing sugarcane production; 30 million liters a year are now blended into gasoline (Hall, 1983, p. 77). Experience with the Kisumu plant in Kenya has been negative, apparently owing to poor planning (World Bank, 1982e, p. 31). The Brazilian program is, however, very large and well-documented. The history of PROALCOOL provides unique illustrations of the complex, "systems" aspects of the introduction of a new type of biofuel—how a number of independent elements of the economic system play essential roles in the process. It also provides an example of a large bioenergy industrial organization that works—and illustrates a case of massive government intervention in the marketplace.

Systems Aspects. Any new energy source is part of a complex total economic system. The success of alcohol projects necessarily depends on the successful association of agricultural systems, alcohol distilleries, and assured markets. Therefore, these institutions must be linked by a reliable network for the collection and delivery of raw material feedstocks and by an alcohol product distribution network (World

4. The word "successful" must be qualified by the fact that the heavy direct government intervention in the program, discussed below in the text, makes it difficult if not impossible to sort out the microeconomic problem, much less the macroeconomic. See for example, Islam and Ramsay, 1982.

Bank, 1980, p. 55). It is difficult to generalize, but it appears that in at least one case (Costa Rica), the absence of the truly systematic approach may have been one contributory cause of failure.[5]

Such strictures about "total systems" may then seem self-evident. Indeed, the real question may be can new commercial bioenergy firms—in this case fuel alcohol producers—be introduced in such a way that the rest of the existing agricultural and economic system will automatically develop support for them, or must the entire system be changed at once? The Brazilian case gives one answer to this question: the entire Brazilian program has been predicated on massive government interventions in all relevant sectors. It is instructive to the bioenergy planner to see—in the following subsection—how that intervention has worked.

To really make a fuel alcohol industrial system work in some countries, planners may have to combine future fuel alcohol programs with the utilization of other by-products such as construction materials from bagasse or plastic from wood feedstock residues. In fact, it may make sense to routinely plan for multi-industry complexes to be set up around bioenergy schemes. Such schemes have been suggested but have not yet been successfully implemented (Ramsay, 1980b, p. 81; Hertzmark et al., 1980, p. 15).

Ownership and Organization: The Case of Brazil. No industry begins in a socioeconomic vacuum. Indeed, there are definite advantages to using existing firms for new bioenergy projects. They can select from a large tried and proven management staff. They have a lower administrative burden, built-in sources of internal funds, as well as established credit ratings with suppliers and ready access to bank loans (SERI, 1981, p. 109). In the event, the alcohol industry in Brazil, now comprising over 300 distilleries, took off from the firm base of an existing large sugar industry—including a cost-competitive domestic engineering capability in conventional sugar technology (World Bank, 1980, p. 32). It seems apparent that the existence of such an industry constituted an important advantage for Brazil, an advantage that few other developing countries will possess to the same degree.[6]

5. One of the first responses of the Costa Rican government to the oil crisis was to construct an annex to an existing sugar factory (CATSA); this distillery was no sooner open than it was immediately closed again on grounds that it was "uneconomic" (U. of Costa Rica, 1981, p. 76). The Costa Rican experience has been a story of many different plans, not all of them necessarily consistent with each other or with national needs. They have included plans for several new private distilleries, together with a relocation of the distillery associated with the national beverage alcohol monopoly, and contracts by the National Council on Productivity with three private companies to produce alcohol from sugar cane or crop residues. These are just a few of the different schemes proposed or partially implemented by different government agencies (see for example, U. of Costa Rica, 1981, pp. 67-68).

6. Not only did the Brazilian program build on an existing network of sugar factories and also distilleries, but even new distilleries especially built for the PROALCOOL program were able to benefit from the existence of national engineering and construction expertise, manufacturers of equipment, and supporting government institutions.

The ownership and organizational aspect of the alcohol program was settled by leaving the responsibility for the agro-industrial projects—the cane growing and fermenting and distilling—to private entrepreneurs. The distribution of alcohol, on the other hand, was assigned to the state-owned petroleum company, Petrobras (World Bank, 1980, pp. 55-56). This public-private division itself has not proved to be hard-and-fast; in the development of new feedstocks, such as manioc, Petrobras has played an active role alongside private industry in direct production (distillation) operations.[7]

This Brazilian case must be viewed as suggestive but not definitive; undoubtedly, other types of industrial organizational schemes deserve consideration by bioenergy planners—total nationalization or total privatization, for example.

Government Intervention in the Marketplace: PROALCOOL. The government role in the PROALCOOL program is overwhelming. First, the government tightly controls the prices not only of alcohol, but of related products: the key price control is not on alcohol at all, but on sugar. The governmental Institute for Sugar and Alcohol sets the prices of both sugar and sugar cane to allow producers a reasonable rate of return (Islam and Ramsay, 1982, p. 11).[8] The price of alcohol received by the producer is then set by the Institute in such a way as to maintain a stable equilibrium between sugar and alcohol production.[9] The price at which alcohol is sold as fuel is set in cooperation with the National Petroleum Council. Finally, the state petroleum monopoly also fixes the market prices of gasoline, giving the government considerable leeway in allotting shares in the fuel market to alcohol and gasoline.

This price fixing structure could not function without a system of production quotas. At the distribution end, the market for all alcohol produced by distilleries is guaranteed at the officially fixed price, but little risk of overproduction results. Production quotas for different types of alcohol as well as all authorization of exports will have already been set by the National Alcohol Commission (Carioca, 1981, pp. 34, 36). Furthermore, the Institute for Sugar and Alcohol sets planned production limits for both sugar and alcohol on a regional and unit basis for each individual mill and distillery. It also controls sugar production indirectly through the Annual Crop Production Plan which

---

7. A pilot manioc plant was built by Petrobas in Minas Gerais, and two other distilleries were reported in the works for the state of Santa Catarina. On the other hand, two private firms also were reported building large (60 thousands liters per day or more) plants using manioc feedstocks (SERI, 1981, pp. 67-68).

8. The prices account also for regional variations. There is also a Sugar Export Fund used to cover year to year fluctuations.

9. The IAA sets the equity price between sugar and alcohol, in agreement with the Ministry of Mines and Energy, recently at one [50 kg] sack of sugar being equivalent to 40 liters of alcohol (Carioca, 1981, pp. 35-36).

helps growers to plant their acreage in cane consistent with the alcohol and refined sugar quotas (Islam and Ramsay, 1982, pp. 10-11).

Finally, the government subsidizes many projects for modernizing and enlarging alcohol distilleries or establishing new ones. Direct suppport is also given for the improvement of alcohol production and utilization technology, and the government supports research on technology for the production of alcohol from new feedstocks (Carioca, 1981, p. 35).

Again, such a complex support program by the government may not be necessary or desirable for the establishment of new bioenergy projects in other countries. In Zimbabwe, it was reported only that the operating firm, the Triangle Sugar Estate, negotiated a fixed price for ethanol with the government (World Bank, 1982d, p. 24). At any rate, it is useful for bioenergy planners to understand the depth and complexity of the planning that has gone into the Brazilian PROALCOOL program.

Other Commercial Bioenergy Projects

There have been many proposals in many countries to undertake various types of bioenergy schemes that could be plausibly classified as "commercially feasible." A number of such types of schemes have complex goals—some noneconomic—and will be discussed in chapter 12. Others are more explicitly community projects, and will be discussed in chapter 11. But it is often not clear which projected large-scale schemes that are planned for commercial viability are in fact likely commercial prospects. In particular, existing plans may not consider adequately—if at all—how the proposed project will achieve market penetration. A typical case might involve the introduction of wood plantations under government auspices, often with outside donor assistance, where the ultimate commercial value of the enterprise remains uncertain. For example, in 1976 and 1977, an FAO and United Nations Development Program (UNDP)-funded forestry project established 1,562 hectares of industrial plantations in Upper Volta as part of a larger program for establishing 3 thousand hectares of new woodlands in the country (Winterbottom, 1980, pp. 25-27). This project should naturally have featured the sale of forest products, including fuelwood, in local markets. The local Forest Service, however, proved to be unable to intervene successfully in the marketing of fuelwood—which had in fact been traditionally tightly controlled by the private sector. In view of other problems that arose in the project, however, there is some question whether such projects represent a specialized failure of "commercialization" through marketing deficiencies or merely a failure of operations overall.[10]

The future might be brighter. One report on a set of bioenergy options considered for Costa Rica, for example, has estimated that governments or donors could stimulate the private sector to undertake a serious program substituting charcoal and other biofuels for imported oil through a few reasonably inexpensive planning and demonstration projects.[11] Gasifiers, in particular, were recommended with the suggestion that the favorable economics of such biofuel substitutions would in time spontaneously encourage private entrepreneurs to make rational changes in energy conversion.[12]

In general, the problem of market penetration, which has been studied at least to some extent for community systems in developing countries, has been somewhat neglected for commercial bioenergy systems. This lack of study leaves a serious gap in data needed for bioenergy planning decisions.

Bioenergy As a By-Product of Commercial Operations

Producing biofuels as a by-product of another commercial operation has obvious advantages. If the by-product is presently considered a waste, or if it is sold at a lower value than if it were sold for fuel, there should be a firm economic impetus toward increasing profits by using it as a bioenergy resource instead. Even if conversion of the by-product to a bioenergy use happens to be involved or complex—requiring costly additions to existing facilities, for example—the mere fact that an organization is involved that has management, labor, and capital already in place and at work will make it much more likely that production and marketing problems will be overcome. Sometimes, indeed, new biofuels can be produced almost as a "new product line," rather than a by-product. That is, the methods of operations of the firm may be changed, but not drastically, in order to accommodate bioenergy production or conversion.

Timber Operations With Energy By-products

One of the great success stories of biomass energy in the United States and in the rest of industrial world is the recent increased use of

10. Apparently the projects suffered from a lack of qualified local personnel and local funding, and operations appear to have headed downhill once FAO technicians, Peace Corps volunteers, and outside project funding disappeared in 1977.

11. See Meta (1982, p. vii) where an additional $0.5-1 million estimated addition to existing assistance program needs was estimated.

12. Meta (1982, p. viii) notes however that a lack of experience with technology such as gasification will make it probable that they will be imported. The stimulation of the manufacture of new conversion equipment was thought to be beyond the unaided competence of the marketplace.

sawmill wastes and of wastes from pulping operations as fuel both for process heat in timber operations and as sources of cogenerated electricity. Such use of timber waste by-products are even a more obvious option in many parts of the developing world where inexpensive alternatives like natural gas have never been available.

Even greater possibilities for the harvesting of larger bioenergy crops by commercial timber firms lie in the concept of combined extraction. Combined extraction means a fuller utilization of the "crop waste" from the timber process by collecting and using it as fuel or for making charcoal. In particular, during operations for sawtimber or pulp, the tops and smaller branches of trees are often left as forest litter, as are whole trees that are too small or misshapen or of otherwise poor quality. These rejected parts can, under favorable circumstances, be harvested and transported along with the wood selected for fiber use, with considerable potential for added profits to the firm (De Beijer, 1979, p. 28).[13,14]

Combined extraction has been a feature of the well-known Jari operation in Brazil. In addition to large amounts produced for lumber and pulp, harvesting and milling residues plus wood cut explicitly for fuel were burned in two steam boilers to drive a 55 MW turbine generator that produced enough electricity for both the Jari industrial complex and for the nearby town of Monte Dourado.[15]

A vital question is: what can governments do in order to assist this process of increased use of timber by-products? This question was addressed briefly in a study for the United States (OTA, 1980, vol. I, p. 150), which suggested that residues from national forests be made available for exploitation—an option that can certainly be adopted in many developing areas. It also recommended the establishment of "concentration yards" for storing residues and fuel-quality woods. Finally, loan guarantees and accelerated depreciation schedules were suggested as useful indirect subsidies. Such policy tools could be of interest for bioenergy planners in developing nations.

Sugar Wastes: Bagasse and Molasses

The residues left over when cane is milled to extract the sugar juice are widely used in the Third World as a source of process heat for sugar refining and distillery operations. Any leftover energy from the

---

13. Depending on policy, bark and leaves can also be harvested or can be left behind as litter in order to help prevent erosion and to provide nutrients. See chapter 5 for environmental caveats.

14. Some of the obvious economic advantages come from a spreading of the cost of capital and fixed expenses: heavy equipment can be shared as well as license fees, survey costs, and administrative expenses.

15. Four thousand hectares per year were needed to supply this electricity. While the scale of the scheme may be larger than most, the concept was reported as not unique in South America (Hartshorn, 1979, p. 12:7).

burning of bagasse can also be used for other purposes such as the generation of electricity. This type of operation can sometimes be carried out by an ordinary sugar firm, with minor adjustments in operations. Other times, the by-product itself can become important enough to warrant changes in organization of production and marketing within the industry.

One successful effort of this kind was mounted by commercial sugar firms in Hawaii. This project involved burning bagasse to supply steam for sugar operations and to operate a 20 MW generator to provide power for the mill operations and excess power that could be sold to local utilities.[16] The sugar operations lasted only nine months, so that bagasse could only be burned during that time. A pelletizing operation was set up to make a stable, dense fuel from bagasse, and during the three off months, bagasse pellets were used to heat the boilers instead (Castberg et al., 1981, p. 112).

Many of the technical and supply problems in this Hawaiian case were by no means trivial and the implementation of this system required the use of outside construction contractors and experts on electric power.[17] From the organizational point of view, however, the company chose to rely on its own employees to design, plan, and carry out its energy program. The management decided that intimate knowledge of the sugar processes, with the ability to integrate the energy operations into them, made for overall efficiency—as opposed to relying on the more specific technical expertise of contractor personnel (Castberg et al., 1981, p. 150).

Another more fundamental organizational step was taken in order to make this project work. A sugar cooperative was formed— the Hilo Coast Processing Company. The cooperative was designed to secure a larger profit for the growers and to encourage a better division of labor, especially for the new energy activities. But most important, the adoption of the cooperative organizational form opened up new outside financing opportunities (Castberg et al., 1981, p. 89).

Making alcohol from molasses, the carbohydrate residue from the refining of sugar, is a common practice in such sugar producing countries as Brazil. The organizational advantage in producing this biofuel from a sugar industry residue is— again—that the same established administrative and technical resources can be used for energy production as for the sugar processing itself (World Bank, 1980, p. 33).

16. Sales of power were reported as $610 thousand, with $315 thousand worth of steam and power used for process heat. The use of bagasse as fuel, rather then for landfill as previously, also saved on waste disposal costs. The costs of the investment were $8,000,512 (Castberg et al., 1981, pp. 87-88).

17. The sugar operations lasted only nine months, so that bagasse could only be burned during that time; during the three off months, pellets were used to heat the boilers instead (Castberg et al., 1981, p. 112).

Biogas

Commercial production of biogas as a primary product has failed to catch on under present energy pricing regimes. But the production of biogas can also often be justified as a by-product of pollution control or merely as an opportunity for making economic use of the carbohydrate portions of the wastes that would have little or no value as fertilizer or soil amendments. In particular, at the Maya Farms in the Philippines, biogas installations have been used for pollution control for a livestock operation. In this installation, 500 thousand cubic feet of gas were produced per day from the wastes of 10 thousand pigs (Terrado, 1981, p. 146). Large-scale units have also been built as part of the extensive Chinese biogas program: large digesters (over 2000 cubic meters capacity) using urban wastes as a feedstock have been reported (Taylor, 1981, p. 210). On the other hand, reports such as those from Papua New Guinea on the current inoperability of plants built to operate on urban wastes at Lae and on coffee wastes at Waghi Mek suggest that this technology may not yet be trouble-free for by-product bioenergy operations (World Bank, 1982h).

Conclusion

Bioenergy production by large commercial firms on its own merits is still a relatively unusual activity, especially if we neglect medium-sized producers of charcoal and fuelwood. The virtues of dynamism, access to capital, and cost-accountability make the large-scale commercial, government, or parastatal bioenergy option appear attractive in principle, however. Therefore, the "semi-modern" case of fuelwood and charcoal production in Hyderabad may be of some interest for other mixed government-private sector operations on a medium or large scale. The isolated case of charcoal utilization in the Brazilian steel industry may serve as a directly useful model for imitation elsewhere. Finally, the Brazilian fuel alcohol scheme admittedly has its unique elements, but its success—at least in achieving levels of production—make it a necessary part of the curriculum of the bioenergy planner.

The outlook for bioenergy as a "by-product" of commercial enterprises appears brighter. Commercial firms should be happy enough to go into new bioenergy outlets for their by-products, as long as the operations are economically feasible. Providing that cost-benefit criteria are satisfied, the entrepreneurial and technical capabilities of commercial firms could be a promising reservoir of talent and capital for furthering bioenergy production.

11

Community-Managed Operations

Introduction: Communities and Hidden Resources

There is a great potential for the introduction of bioenergy in the Third World through projects operated by the community acting as a whole. We have seen in chapter 10 that important strides can be made in bioenergy through commercial ventures. These ventures have the advantage of the profit motivation and the ability to use established firms with existing skills, equipment, and access to capital. Entrepreneurship is often lacking or capital is unavailable, however. Or projects may produce potential benefits like environmental protection for the community or the nation as a whole that do not directly help the individual businessman. The costs of the biomass project in ordinary marketplace terms may therefore appear to be high. One way to lower those costs is to uncover "hidden resources." These hidden resources may be the underutilized labor of much of the rural population, land areas that are yielding poor or no economic returns, or unused management skills. If costs are still too high, a degree of subsidization from national or provincial governments may also be justifiable in recognition of possible social or environmental benefits to the nation as a whole. Indeed, community management might be especially appropriate when the benefits go beyond the production of fuels, extending into environmental protection and other collective needs. In particular, social goals such as income redistribution and the stimulation of new community institutions and the development of community leadership, as well as political goals related to social unrest, may be reasons for going the community route.

Hidden resources can be tapped at the level of the individual farmer or villager or at the community level. In chapter 12, we will examine how schemes involving farmers or other smallholders can be organized. In this chapter we examine the problems and possibilities of community-operated projects. Although the majority of these projects are usually targeted to rural villages they may also be appropriate for some urban communities.[1]

Community-operated projects can, in theory, encompass a whole range of biomass energy projects, which themselves are often integrated with other rural development purposes. In practice, most community schemes have focused on fuelwood and charcoal. Nevertheless, as we have seen in chapter 10, a large number of biogas plants have been implemented at the community level in China, for example.[2] Similarly, the Indian government has been promoting the development of community-scale biogas operations (KVIC, 1983, p. 1). Furthermore, there is no reason why community management should not be considered for schemes involving the production of producer gas, the direct combustion of wood to produce electricity, or any other type of bioenergy technology.

In this chapter we discuss the nature of community-operated projects at the national, regional, and community levels. Under the national and regional organization levels, questions arise on funding, publicity (propaganda), and the role of extension services. At the local level village leadership (i.e., management), property arrangements, division of labor, salaries, prices, distribution of the bioenergy product, and liaison with other institutions, like funding agencies or forestry departments are of interest. Possible constraints on the success of such community-centered projects are numerous, and such issues as scale, land use, and legal problems must be examined by the planner and participating members of the community.

Organization of Community-Managed Operations

Because these community-managed projects are customarily oriented to meet the needs, both energy and societal, of people in the community, the question of how organizations relate to people—and people to organizations—is paramount. Any project implies human as well as physical and technical change. How the organization (or leaders) that bring the project to a community are perceived locally will inevitably affect the acceptance level of the project (World Bank, 1980, p. 31). Do the project leaders have credibility within the community? Where does legitimate power lie? What type of organizational

1. For example, a number of projects in Latin America (Thrupp, 1981, passim) have been developed with the participation of municipalities and development associations as "social benefit" projects involving prisoners and alcoholics as laborers in establishing recreation areas and school nursery projects.

2. Most Chinese community digesters, from an organizational view, constitute subsidiary activities of piggeries or of schools, factories, hospitals, and other institutions (Taylor, 1981, p. 209). Whether these schemes in centrally-planned economies are called "community-managed" or "national-commercial" is somewhat arbitrary: we have treated them both here and in chapter 10.

framework works best? These are the kinds of questions that must be addressed.

Organization at the National and Regional Level

The average community bioenergy project cannot do without some help from the national or regional governments—or from outside donors. At a minimum, typical forestry projects receive aid in the form of seedlings and technical advice. Quite often, the national government or its agencies are very closely involved, with or without the help of donors, as with the entrepreneurial role played by one agency (Khadi and Village Industries Commission) in the community biogas program in India (KVIC, 1983, p. 4).

The national government may intervene in the community to stimulate bioenergy programs as part of its interest in other national goals. For example, a parastatal group in Senegal sponsored nurseries for nitrogen-fixing trees as part of a program to increase peanut production by preserving soil from leaching (Hoskins, 1979, p. 24). Often, the national government intervenes on behalf of its vested interest in local institution building.[3] At times, the interaction of a bioenergy project with the national government can become quite complex, and community projects can share in the opportunities of and problems posed by extensive government media campaigns for fuelwood or biogas production. These complexities of interactions are discussed at more length in chapter 14, where many features common to all kinds of biomass projects—and not only community projects—are discussed. Here we mention only some of the national government roles that may be important in bioenergy planning for community-managed facilities.

This government aid can be directly financial, subsidizing the community project through loans or grants, or indirectly so, through financing the work of government agencies offering technical assistance. Sometimes, of course, this financial aid will come from a donor, rather than from the government; in any case, the importance of securing a long-term commitment of funding, especially for forestry projects, has been often emphasized by experts (Winterbottom, 1980, p. 41).

A minor but not insignificant role can be played by the government in providing incentives. For example, cash prizes, ranging from 300

3. For example, in the Burkina Faso (Upper Volta) Project under the Government's Forestry Administration, institution-building (e.g., setting up marketing cooperatives) accounts for 57 percent of total project cost. A Malawi project emphasizes the strengthening of the government's renewable energy planning capability (World Bank, 1980, p. 3).

to 1,000 rupees (about $30-$100) have been awarded annually to the best community woodlots in the Indian state of Gujarat (Java, 1983, p. 6); in Ecuador, a system of requiring community contributions of labor paid for with a special food ration has been tested (Thrupp, 1981, p. 34).

By far the most common government contribution to woodlot projects, however, is assistance from the forestry department. For example, in the Gujarat program, the state forestry department not only provides the seedlings to the local people but it also promotes the project by encouraging the villagers to set aside the land necessary, but the forestry department staff also supervise the planting and tending of the trees.

The design of extension programs must reflect specific local needs. Where tree growing is a well-known art, extension eforts at forestry education may be misspent, as in some cases in Tanzania (Skutsch, 1983, p. 50). In other places, close forestry department supervision may be essential. In most places, however, low quality of seedlings and, especially, late delivery can be fatal to chances of success; these problems were listed as key cause of failure of village plantations in Burkina Faso (Upper Volta) (Winterbottom, 1980, p. 21).[4]

Whatever the form of aid from the national and regional governments, it has been speculated that too much aid can be as ineffective as too little, leading in some cases to paternalism and an excessive dependency of the villagers on the government (Araya, 1981, p. ii).[5] Presumably, an emphasis on the development of community hidden resources will lessen this danger.

Organization at the Community Level

There is no one best way to organize a community-managed bioenergy project. It is obviously more difficult, however, to carry out such schemes where villages have little tradition of successful community cooperation (Parker, 1973, p. 4). For example, in China and other centrally-planned economies, the project will normally fit into the usual collectivized scheme. (Taylor, 1981, p. 105ff) (For example, reports that the political arm of the system now bears ultimate responsibility for all patterns of rural development and the resulting holistic approach to such development.) In Malawi, rural and urban councils that were already in existence organized plantations to produce fuel-

4. The timing can also be bad, with extension services provided heavily before the project, but with insufficient follow-up (Winterbottom, 1980, p. 41).

5. Such symptoms developed in the Tanzanian program where the increased active involvement of the Forest Division in the actual planting (in 37 percent of the programs) may have distorted its original role as a project catalyst.

wood and poles (World Bank, 1980a, pp. 15,16). In Indian communities the standard village council (*panchayat*) is usually the relevant institution. In one of the few successful collective biogas projects in India, however, the plant is located in a religious community (*ashram*) and the leader of the *ashram* makes the decisions (Bahadur, 1984). In forestry projects in the Sudan, the farmers' union was initially the sole representative of the villagers; however, in order to allow women to participate, special program committees at the village level had to be established (Hammer, 1982, p. 55). The Tanzanian woodlot program took advantage of a new emphasis placed on the political role of village councils as part of a radical national program of compulsory resettlement of scattered rural populations into relatively large (250-600 family) settlements (Skutsch, 1983, p. 4).

One organizational imperative that would appear evident, but has been often neglected in practice, involves the question of education and training. If the project is to be a community effort, the members of the community will work most efficiently if they feel that they are participating actively, are able to contribute their ideas, and are being informed in an effective manner of the way the project is to work. The failure to achieve community participation has been widely noted as a cause of failure in Niger, especially in land tenure problems (see below) (Noronha, 1980, p. 19). Specific difficulties can also arise when villagers fail to understand—or believe—that they will share in profits, as reported, for example, in cases in Burkina Faso (Upper Volta) (Hoskins, 1979, p. 7).

Organizing protection schemes for village woodlots is also an important function. In cases in China, protection from poaching and grazing was organized by staff from a local "brigade" organization (Taylor, 1981, p. 134). In Korea, rotating two-man patrols from the local Village Forestry Association report illegal cutting, pests, and other problems (Ahn, 1978, p. 8).

A key organizational challenge is the timing and division of returns from the project. Many biomass projects (especially fuelwood projects) are long-term in nature. How do the villagers survive while the forest matures? In many cases, the villagers may be able to support themselves through agriculture and other activities during the years before the first harvest. Another possibility is to give the villagers subsidized loans during the growth period, which they repay after they start cutting.[6]

6. One type of scheme is the concept of fuelwood "futures" in which a payment for the living trees could be considered an advance toward the future products (similar in concept to grain futures). When this type of payment was adopted in Senegal (Hoskins, 1979, p. 45), the survival rate of seedlings shot up to almost 100 percent as compared to a 20 percent survival rate in schemes where there was only a payment for planting.

How are the profits to be divided? The Indian Gujarat Project has set up several schemes to allow for differences in the degree of involvement of the villagers. About 25 to 30 percent of the villages fall under the "Self-Help Scheme" in which the villagers contribute labor or its equivalent (cost of land, preparation, planting, maintenance, protection, etc.); all the benefits go to the village (Ranganathan, 1981, p. 11). In contrast, under the "Supervised Village Scheme" all costs have been borne by the Forest Department. The direct cost of plantation and management fees, and 25 percent of development costs are to be deducted from revenue and the rest to be given to the villagers. The forestry program in Korea set up a fairly rigid schedule of the division of profits between owners, planters, harvesters, and others.[7]

One Indian plan for collective biogas plants is to give away the gas free and then to divide the slurry residue among the community members in proportion to the animal wastes contributed (Bahadur, 1984). Unfortunately, distribution systems have not been free of equity difficulties even under socialism; in some Chinese projects individuals were reported to have been inadequately compensated for the proportion of waste that they had contributed (Taylor, 1981, p. 208).

In some areas of the world, community schemes have even benefitted from financial self-help. Villages in Korea have set up common funds from voluntary contributions to assure the continuing financial viability of the project (FAO, 1982, p. 84).

Constraints on Community-Managed Schemes

Since a community-managed scheme is usually a kind of joint venture between the government or a donor and a community, the usual discipline of the marketplace is significantly reduced because additional outside support can often compensate for local failures. Therefore the planner cannot rely on purely economic indicators like profit and loss, and he must consider carefully a number of factors that can be constraints on the project.

The question of choosing the size or scale of the project can be a problem, as well as determining the scope—how many different energy

7. For the Korean program, the national government set up the following profit-sharing scheme (FAO, 1982a, p. 48).

Classification	Crop-tree	Timber Forest	Fuelwood Forest	Natural Forest
Forest land-owner	10%	10%	10%	10%
Planter	15	30	40	—
Forest Manager	75	40	30	60
Timber Harvester	—	20	20	30

and non-energy components should be included? The project location and the technology to be used can present difficulties. Personnel problems—strategies for staff and management—are of constant concern, and an array of legal, social, and cultural problems can raise "non-economic" difficulties.

Project Scale and Scope

What are the appropriate size (scale) and scope of a community-managed bioenergy project? Projects that promise to produce large amounts of energy may be preferable to those producing small amounts. It is also easier to organize and implement one large project rather than many smaller projects. On the other hand, many an ambitious community scheme has withered on the vine either because of its sheer size or because of its complexity. If a small-scale modest project does turn out to be more desirable at the community level, then how many villages should be included in each project? In answering this question, one must be aware that the effectiveness of a particular scheme means little from the perspective of the regional or national planner if it is not expanded to other villages[8] (Barnes et al., 1982, p. 42). In terms of long-range energy planning, reforestation and afforestation efforts seem to be most successful when the social forestry program donors avoid the temptation to plant as many trees as quickly as possible, rather than paying attention to *how* the trees are to be planted, tended, and harvested (Winterbottom, 1980, p. 26)[9]. Furthermore, an integrative approach to social forestry programs is more likely to address the problems of local participation and institution-building. For example, the Gujarat project in western India began modestly in 1974 by establishing village woodlots on grazing lands with free seedlings distributed by the state forestry department. Only after the success of numerous four-hectare village woodlots under this program was a more ambitious Community Forest Project launched in 1980. The stated aims of the project were to: (1) meet fuelwood requirements, (2) secure the participation of villagers through revenue sharing schemes, (3) provide employment, (4) stabilize the environment, and (5) introduce fuel-

8. For example, a small-scale fuelwood project was started in the south of Benin. The seedlings were distributed by the regional development organization and the extension work was done by regional groups. Subsequently, cost-effective production units grew up all over the country, with help from an effective a publicity campaign, and small private nurseries have popped up in the south (Wourou and Tran, 1982, p. 7). (See also the discussion below in chapter 14.)

9. Such efforts have been stressed in Korea, China and India, the three most successful community reforestation projects, and had national impact.

saving measures (more efficient stoves and crematoria).[10] The original program had planted 17,034 hectares of woodlots in 2 thousand villages (out of 18,275 in the state); the new program had scheduled an additional 36,440 hectares (Java, 1983, pp. 8, 10). Other such multipurpose projects have been equally successful in China and Korea: 208 thousand hectares were planted during the first 10-year plan in Korea (FAO, 1982, p. 68). The "lessons" of these exemplary projects may not always be applicable to the develop ment of new projects in a different socioeconomic context. It is notable, however, that they did all build on established bases of smaller projects and did secure the strong support of the villagers.

Location and Technology

Woodlot programs should logically be targeted first to areas where there is a real (actual or predictable) shortage of fuelwood and where the local inhabitants are aware of this need.[11] Lack of perceived need was a key factor in several villages in Tanzania where woodlot programs failed because of lack of follow-up after initial planting (Skutsch, 1983, pp. 47-48).

A fuelwood planting site, theoretically, should be located as near as possible to the point of end use to minimize transportation time and

10. A five-year investment of Rs. 680 (1 rupee equals about 10 cents) million was planned, one-half of which to be funded by the World Bank and the rest by the state government: it includes the following tables of activities (Java, 1983, p. 10):

Activity	Target
Plantations:	
Roadside	31,600 ha
Canal Banks	2,000 ha
Railway Sides	3,400 ha
Village Woodlots:	
Irrigated	2,880 ha
Rain fed	34,560 ha
Forest Areas:	
Reafforestation of degraded forests	30,000 ha
Tree Farms:	
Plantations in privately owned lands	1,000 ha
Total Land Area	105,440 hectares
Seedlings for Farm Forestry:	150 million
Wood-Saving Devices:	
Improved crematoria	1,000
Smokeless stoves (chula)	10,000
Support Activities:	
Research	
Training	
Service Personnel	
Farmer Training	
Publicity	
Communication	

11. Some fishing villages in Tanzania began to have a severe fuelwood crisis when they did not have fuel available to smoke their fish. Manpower had to be increased to collect wood from areas further and further away from the village (Araya, 1981, p. 14).

expense. In practice, however, land use constraints are usually a more influential location criterion for woodlots. For example, where there is close competition between food crops and forestry, the outlook for most types of community forest schemes is bleak (World Bank, 1980a, pp. 18-35). In such cases, as suggested below, there often exists the possibility of organizing food and fuel schemes that are complementary, however.

Land use competition may pose hidden dangers, especially where land tenure (see below) is not formalized. A project in Niger, for example, located a woodlot on land that had traditionally been devoted to grazing. Villagers either pulled up the saplings or tore down the fences to let their animals graze (Noronha, 1980, p. 7; Hoskins, 1982, p. 8).

For other types of bioenergy such as biogas, locations will have to be chosen so as to be within reasonable pipeline distance of the users. Long pipelines require energy for pumping and are not economic for raw biogas, which has a low specific heating value that makes transport costs per energy unit relatively high.

For bioenergy community forestry projects the choice of species is a complex question (see chapter 4)—and also a point of contrast with commercial timber operations. In conventional commercial logging operations, the choice of tree species has been based upon the "workability," color, and strength of the wood. Energy species can be selected for energy value, growth characteristics, or possibilities for combining fruit, fiber, and fuelwood crops. Multipurpose species, for example, are especially important for community operations in which a variety of individual interests have to be accommodated.[12] In this regard, agroforestry schemes can provide a sophisticated means of meeting multiple use goals. In Gujarat, a three-tiered system of cultivation is being promoted: a grass or other fodder crop in the lowest canopy, agricultural crops in the middle canopy, and biomass for energy in the topmost canopy (Java, 1984, p. 6).

Staffing and Management

The personnel chosen to carry out a community-level project is a critical factor in the success of the project. The government or donor project personnel must have at least a minimum of credibility with the community. For forestry projects, one of the traps one may have to avoid is that of using foresters in those areas where they are regarded as policemen—guarding government lands from poachers—as opposed

12. Another way to do this, for example, is to introduce to cattle-raising communities simple biogas digesters to produce methane for space heating and cooking (See chapter 6).

to agents who are there to serve as a source of information, expertise, and technical assistance. Specific personnel choices, however, will depend on local situations. A Tanzanian analysis (Araya, 1981, pp. 63-64) suggests the following as the core personnel for a village forestry program: one administrator; two graduate foresters advising on the choice of species, siting, planting, soil conservation techniques, and harvesting methods; one social worker/sociologist; and priority assistance as needed from the resource assessment section of the forest division.[13] Although this list gives an idea of the different kinds of skills needed in community bioenergy projects, it would be impossibly demanding in some contexts. In practice one has to, and can, make do with less stringent personnel requirements.

One of the most important points is that the local community leaders who manage the project must be trusted. In the case of community system the management is usually a village council or headman, and it has been noted that a lack of trust of village leaders—especially in their role as distributors of biofuel products or revenues—has been one of the primary causes of failures in some Tanzanian villages (Skutsch, 1983, p. 50). Continuity of village leadership can also be a problem. In a case in Peru, village leaders changed every two years, severely handicapping a forestry project (Parker, 1983, p. 4).

Successful project staffing at the working level may have to take into account cultural constraints. In a project in Senegal, tasks were divided along gender lines; the men planted the seedlings while the women carried the water for irrigation (Hoskins, 1979, p. 22). Furthermore, it has been observed that if the fuelwood crisis is reaching such proportions that the men must take up more of their time to gather fuelwood—traditionally a woman's task in many areas—village forestry will quickly get more priority.[14]

On the positive side, the strong commitment of women can often help assure the success of a community-level project. In a great many rural villages, women have the role of fuelwood gatherers. It is therefore essential to gain the commitment of the women if the culture dictates that they will be doing most of the labor related to biomass projects. In discussing Africa, one investigator said: "Women are the primary users of fuelwood and their perceptions of the advantages offered by social forestry projects may hold the key to community acceptance" (Hoskins, 1979, p. 12). (See Appendix 12-A for more on women in bioenergy development).

---

13. One forestry agent reportedly cannot be expected to initiate, supervise and monitor a planting program of over 60 hectares per year unassisted (Araya, 1981, p. 13).

14. Araya, in describing a forestry project in Tanzanian fishing villages, cites such a case, where the mobilization of the village to accept afforestation was guaranteed by the decisionmakers' (men's) awareness of the burden of firewood collection (Araya, 1981, p. 14).

The planner may find one of the greatest roadblocks to the success of community schemes is the absence of established management skills or experience. The KVIC in India has built some 30 "community" biogas plants, with 90 more in the planning stages (Chopra, 1984). All of these plants are as yet still run by KVIC personnel, even though they are supplying gas and slurry to local communities. The stated intention is to transfer the management after three years to the "project committee" set up by community at a later stage, but no such transfer has yet taken place (KVIC, 1983b, p. 4). Furthermore, the aspects of the Gujarat forestry program featuring self-help have been reported as relatively unsuccessful compared to those managed by the Forestry Department (Skutsch, 1984). Training local inhabitants in management skills will be a priority item on the agenda for many bioenergy projects.

Labor requirements for the project must work within the time constraints of the villagers' other tasks. Even if payment for labor were to be offered—for planting seedlings, for example—there is no guarantee that labor would be forthcoming if it interferes with other, higher-priority labor needs like agriculture. The timing of planting and harvesting is crucial, both as a technical decision to be made by the forestry specialist and as a social decision to be made by the villagers as to when time is "free"—e.g., off-season from harvesting, planting, or fishing (World Bank, 1980d, p. 16). In one case in Senegal, villagers reportedly destroyed the trees so that they would not have to leave their work in the fields (Hoskins, 1979, p. 7).

Legal Problems: Land Tenure

Legal constraints on land use can be even more important than labor in the village woodlots. For instance, in Tanzania the establishment of communal woodlots of two to five acres often required the redesignation of parcels of land from private to common use, with the obvious potential for social problems (Skutsch, 1983, p. 31).[15]

Differences between formal and nonformal systems for land use can also be significant. The colonial legal structure that is still the basis for land law in many developing countries often constitutes only a formal patina on actual land use practices, such as customary law of tenure and inheritance (World Bank, 1980d, p. 16). For example, trees may be defined as the private property of a community, subcom-

15. A concrete example of how these legal differences can be compromised can be found in the Korean case. Seventy-three percent of the forests are privately owned. Village associations plant, maintain and harvest trees on woodlots of private forests and the landowner receives only 10 percent of the profits (See footnote 7 above). In contrast, 10 percent of total land in India is government-owned forest. But the *panchayats* (village councils) are allowed to assume most responsibilities for this land by the central government (World Bank, 1980b, p. 3).

munity, or individual; the rights of each of these parties must be sorted out to keep ambiguity from being an obstacle to participation (Barnes et al., 1982, p. 14). "Modernization" of laws made add to the confusion. Since private land was nationalized in 1963 in Tanzania, farmers using "state-owned" land may, on the one hand, hesitate to plant for the future. On the other hand, they may also be constrained by their awareness of earlier traditional rights (Noronha, 1980, p. 17).

In some parts of Africa, indeed, trees are unowned or unregulated property; anyone who reforests might have no right to harvest the tree crop resulting from his own actions, causing a lack of incentive. In any event, tree growers might have to deal with patterns of local "rights of use," allowing nearby residents to collect minor forest produce, hunt animals, dig wells, etc. (Adeyoju, 1976, p. 46).

In the case of government ownership of forests, illegal cutting and a lack of effective enforcement of poaching laws are widespread problems. In many countries, squatters pose a severe problem for land use changes. "Regularization" of their status such as has been planned in parts of India (Madhya Pradesh), will probably provide only modest assistance, if any, for future bioenergy planning, however (Campbell, 1980, Annex 1, p. 2). A corresponding phenomenon in Pakistan, but featuring squatting by the rich rather than the poor, involved a process of gradual "privatization" of government and communal lands through collective apportionment of usufruct and the conversion of a pattern of customary usage into ownership (Cernea, 1980, pp. 17-18).

Conclusion

Community-managed bioenergy projects inevitably cause extra problems compared to commercial projects or to projects utilizing individual farmers. But the community route can also lower costs by making use of underutilized labor and land resources and can also bring extra benefits through the development of community institutions. Although the community-managed project is not necessarily suitable for all social milieus, it may be exceedingly effective in making possible breakthroughs in otherwise stagnant socioeconomic environments.

In order to take advantage of this option, special care must be paid to factors such as management, a fair division of labor and revenue, conflicts and ambiguities in land ownership, and other economic, social, and cultural problems.[16]

16. This chapter is based on drafts by Adela Bolet.

APPENDIX 11-A

A Resource Lost: Barriers to the Participation of Women in Bioenergy Projects

by Susan Piarulli

Introduction

Mnzava (1980, cited in Skutsch, 1983, p. 2) observes that women in some parts of Tanzania report spending 1600 to 2400 hours per household per year—that is, two to three *months*— collecting wood. It is reported that for some, more than time is lost as a result:

> In Gagna [Mali], Syn and Goundaka women pointed out that the physical trauma resulting from walking 15-17 kilometers *one way* often renders them incapable of lactating or performing any strenuous household tasks such as drawing water or pounding millet for two or three days (Ware, 1980, cited in Shaikh and Larson, p. 2; emphasis in the original).

In other regions, where fuel supplies can be obtained only if purchased because local resources are inadequate, women have been known to sell the food that should have gone to feed themselves and their children in order to get the money they needed to buy kerosene, charcoal, or wood (Reynolds, 1975, p. 35). For these women, the situation is near-critical, so it is understandable that planners take their participation in projects designed to conserve or expand fuel supplies for granted. They are frequently disappointed when women fail to act in ways that seem to be clearly in their interest, however. Some reasons that women do not become involved, and some suggestions for overcoming the problems they face are reviewed in this appendix. (For a more thorough treatment see Piarulli, 1984.)

Defining the Barriers

Barriers to the involvement of women may be physical in nature, or manifest themselves in the legal, traditional, or customary practice of a region. The cultural biases of planners and those of organizers in the country may cause solutions, or indeed the problems themselves, to be overlooked.

Physical Barriers

The first classification is the most easily recognized: it is the lack of materials needed to take part in a project. Transportation, for example, may be unavailable to indigent women, who will therefore be unable to get to far-off woodlots without spending an inordinate amount of time in transit. If biomass is to be grown on the homestead, women may lack the financial means to buy seeds, seedlings, tools, etc. Water and fertilizer may be scarce in some areas or already committed to other uses. Women willing to work may thus be overlooked for want of the material means to participate.

Barriers in Law

Barriers in law are the next step up on the scale of difficulty of problems needing to be resolved. These include land-tenure laws, water rights, and criteria for judicial status. For example, in Tanzania—as elsewhere—it is common practice to treat women as legal minors. Like children, they must have male kinsmen represent them in court (Reynolds, 1975, p. 5). This can be an overwhelming obstacle to women trying to assert their rights against their own kinsmen for property belonging to the family. If women would like to secure land, water, or other materials in order to take part in a project, or commit resources already in familial possession, her husband's objection alone may be enough to stop her.

Some land-tenure laws are similarly biased against women. In the Kordofan region of the Sudan, local land ownership was determined by practices of customary inheritance law which allocates to women only half the land ceded to men (Hammer, 1982, pp. 4-5). When the government decided to limit the partitioning of farmland into many small plots, the women received no land at all (Hammer, 1982, p. 26). In cases in Tanzania, the women were observed to be even less fortunate. Upon the death of the husband, all property—farm produce, house, cattle, and land— revert to his eldest surviving brother. The wife (or wives) retains only the property that she brought to the marriage (Reynolds, 1975, p. 18). This inability to acquire or retain land will exclude women from primary participation in bioenergy projects requiring the villagers to contribute their own real estate. Even if men are willing to use the land they control for energy schemes, their coincident control over the disposal of the crops produced may discourage women from taking part in a project absent a guarantee they will benefit from the additional work burden.

In East Africa, for example, women refused to reduce the acreage devoted to subsistence agriculture in favor of cash crops, for men control the returns from the latter while women control the former (Reynolds, 1975, p. 14).

Prejudicial legal practices may also directly affect the distribution of benefits. In one case in Upper Volta, a title was granted to residents who contracted to cultivate land formerly used by local landlords. After planting and nurturing a lucrative garden, some women lost their investment when local tribal leaders "reclaimed" the plot (although the women were "compensated"—with land further from the village) (Hoskins, 1979a, p. 22).

Traditions and Customs

Apart from the legalized practices mentioned above, there are those not codified that inhibit feminine participation. In any culture, there is a network of traditions and customs that dictate what is acceptable social behavior and what is not. In Burkina Faso (Upper Volta), for example, it is inappropriate for a woman to voice her opinion if a man is present (who can voice his for her) (Hoskins, 1979a, pp. 14-15). In one case, men talked of an impending fuelwood shortage, but it was only after planners spoke privately to the wives that they learned there had actually been *no* wood for several months. The women had substituted dung for wood when cooking, and the absence of fuelwood went entirely unnoticed by men (Hoskins, 1979a, p. 11). In parts of the Sudan, the extent of women's travel from the homestead is restricted by Islamic custom, making participation in bioenergy projects far from home limited if not impossible (Hammer, 1982, p. 8).

Traditions and customs in other regions will similarly affect the ability of women to take part in a project. They are usually difficult to discern because they are not explicitly stated; they are simply "the way things are".

Attitudes and Prejudices

The condition most threatening to the potential inclusion of women exists when the planners themselves discount the contribution women can make. Projects are often designed according to Western models. This means that male foresters are trained to deal with farmers of the target region, also presumed to be male. "The primary assumption is that men are the farmers, and if they are not, they should be" (Reynolds, 1975, p. 27). It is often *women*, however, who perform vital support services needed to keep a farm going, a fact frequently overlooked. In one West African country, extension agents showed men the proper depth to plant coffee beans, but the plants continued to die in shallow holes from bent taproots. These agents, caught in the misconception that because men controlled the proceeds from the crop they also did the work, failed to notice that women, not men, dug the holes (Carr, 1978, p. 27).

This case also demonstrates the impact of a negative attitude held by the intended beneficiaries. First, the assumption that important information will be passed on is not supported by the facts in this instance, possibly because the work of women was so taken for granted that men had no interest in instructing them. Second, women may have remained uninformed because they were too intimidated to bring the problems they encountered in the field to light and thereby to find a solution. Women in male-dominanted societies may be similarly unwilling to complain about inadequate fuel supplies, leading to an inaccurate definition of local needs (see below 11-B.2.3). Thus, the subjective evaluation of local men may affect the communication of information regarding the project and the decision-making process that will determine whether the proposal is "worthy" of their or their wives' support, while local women, unwilling to raise their voices concerning issues that are "for men to decide," may miss a valuable opportunity to enhance fuel supplies and improve their quality of life.

Overcoming the Barriers

Various approaches to overcoming barriers to womens' participation have been tried. Some have been successful.

Physical

The very involvement of women may serve as a means for the planner to increase his/her total resource pool to provide the financial and material support women may need. Several international and private organizations have expressed interest in programs that make a sincere effort to include women. Many members of the Private Agencies for International Development, a consortium of 200 nonprofit organizations, have established policies for promoting women's active involvement in their programs and for providing the necessary financing. Private institutions, among them Avon Products, Inc., Chase Manhattan Bank, Exxon, Mobil Oil, Morgan Guaranty Trust Company, and Pfizer, have made available new or increased funding in the interest of exploring women's participation in the informal or formal economic sectors (Smith, 1984, pp. 6-7). Women in Development, an office of AID, is expressively commited to "integrating women into all of its development programs" (McPherson, 1984, p. 1) and has provided technical assistance to missions and AID offices and bureaus to help accomplish these goals as well as spending over $7 million (1981-1983) to support new projects that specifically benefit women and more than $120 million to integrate women into on-going agency projects (McPherson, 1984, p. 3).

Furthermore, the materials needed may be available through local institutions. Kenya's Village Polytechnics, now in over 200 rural communities, produce low-cost goods such as agricultural implements, wheelbarrows, and stoves for sale to women who could not otherwise afford them (Carr, 1978, p. 88).

Legal

Legally binding contracts that include both husband and wife, but which delineate what will ultimately go to whom are one possible solution to problems resulting from property laws. Direct contracts between women and administrators adjudicating possession of resources (such as land) for the life of the project is another. Reserving a defined percentage of resources and harvest exclusively for women is a third.

Local women's groups, well-versed in organizing and directing women despite the constraints of the legal system, may be able to offer suggestions for organizing women to circumvent prejudicial practices. In Kenya, for example, influential formal and informal groups (like voluntary self-help workteams) number in the thousands with more than a half a million members and succeed in involving women in areas previously closed to them (Thrupp, 1983, p. 21).

If the government has a progressive attitude towards women's rights, civil servants may already be in the process of trying to change biased laws, but in many cases the solution will have to come from internal project design. Because project planners usually establish the criteria for participation, carefully defining these criteria may provide the opportunity for women to become involved, regardless of the negative manifestations of land-tenure laws. For example, a project in the Sudan to slow desertification and improve the carrying capacity of the land by planting *Acacia senegal* was redesigned to allow for intercropping. Thus, women, with smaller land holdings than men, could then participate and still have food to feed their families.

Traditions and Customs

Overcoming these problems requires first that they be recognized as practices particular to a culture. Then mechanisms that sidestep customary practices without violating the sensitivities of the people may be found. For example, although it was unacceptable for Voltaic women to be interviewed by male foresters absent the company of a male kinsman, female foresters were able to solicit the women's opinions without interference (Hoskins, 1979a, pp. 14-15). Thus, where women cannot be contacted by men, one solution is to make it possible for them to be contacted by women. Furthermore, to circumvent the indifference with which women's opinions may be habitually regarded, the suggestions made to female extension agents could be routed through male extension agents who then may present them to village administration.

A project in Labgar, Senegal, demonstrates the positive effects of involving women in extension services. A parastatal group (SODEVA) sought to increase the carrying capacity of the soil, and therefore planned to plant trees. Unfortunately, there soon developed a shortage of seedlings, and it looked as though the idea would have to be abandoned. SODEVA officials turned to women for a solution. A female extension agent and a male forester contacted various women's groups with a proposal to establish individual nursuries. Women of Ngodiba, Senegal, accepted the challenge, planting and tending trees locally. The result: not only did the original project succeed, but the women found the nurseries so profitable that they planned to perpetuate the project after its initial stage (Hoskins, 1979, p. 24). Thus, although the definition of proper social behavior seemed to prohibit the consultation of women, ingenious planners found an acceptable way around the structure of custom, to the benefit of their project.

In Kenya, where women had not traditionally been involved in such projects (much less initiated them) a women's group, the National Council of Women (N.C.W.) has directed and successfully maintained a tree-planting program, as well as facilitating the dissemination of cookstoves designed to save fuelwood (Thrupp, 1983, p. 24). The tree-planting effort is especially well-organized and documented. As of 1982, there were 200 N.C.W. "green belts" in Kenya, and the movement shows promising signs of expansion (Thrupp, 1983, p. 25).

Attitudes and Prejudices

In a Sudanese case, *sheikhs* (landlords) at first opposed feminine participation in an agroforestry project. Eventually this opposition was overcome, however: "the experience is that informed discussion can lead to a change in attitude" (Hammer, 1982, p. 65). If men persistently refuse to participate or contribute resources to a project benefiting women, however, several avenues are open. First, women may be given the opportunity and resources to work if men refuse to do so. Alternately, the project promotion could be couched in such terms that men see the benefit to women as an advantage to themselves. For example, if collective family financial resources are used to purchase fuel, it may be pointed out that a plan to make energy locally available frees money controlled by both men and women. In regions where women spend a great amount of time travelling to distant forests, men may be inspired to support a project if they realize that this time is not as productive as it could be. An example of compelling economic logic is found in Senegal, where up to 25 percent of the annual millet crop is lost to rodents and insect infestation. Threshing is believed to reduce this kind of crop loss, but women, who thresh and grind the grain, find little time and energy to do this after spending up to five hours of the day searching for firewood. With a savings averaging $75 per hectare, millet preservation can directly improve a family's financial position if women have more time to complete the necessary processing (Brookhaven, 1980, pp. 81-93).

A final option is to integrate into the project other goals that directly appeal to men. "Integration" is the watchword here: if gains are realizable by selective, limited involvement, men may contribute only to those aspects of the project from which they receive returns.

Even when male attitudes do not stand in the way, a woman's negative self-image may cause her to hesitate to assume a position of responsibility in areas previously reserved for men. The understanding that a place has been created specifically for women within the proposal, however, can give her the confidence to step forward. In Umm Ruwaba, Sudan, late delivery of seedlings for a forestry project led to distribution on a first-come-first-served basis, which penalized women who were, in general, slower to respond. But: "A number of females said they would try to get some seedlings next year, if available, since they had finally realized that women were meant to be part of the target group" (Hammer, 1982, p. 57).

Summary

Numerous barriers exist that impede the participation of women in bioenergy projects—as in other development programs. Overcoming these physical, legal, traditional, and attitudinal obstacles is important because women have an especially important role to play, not only as users of household energy, but also as important economic actors in the local economy. Although bad examples exist, there are cases of creative, successful approaches toward removing these barriers that can serve as guides to planning.

12

Miscellaneous Organizational Strategies: Governments, Donors, Associations, Smallholders, Institutions

Introduction: Any Approach That Works

Anything and everything has been tried in efforts to conserve and upgrade the biomass resource, and therefore there exist projects utilizing a myriad of organizational possibilities—many of which could be of use to future bioenergy planners. In fact, the vast majority of bioenergy projects in the Third World are neither medium nor large-scale commercial ventures (chapter 10) nor are they community-managed efforts (chapter 11). Probably the most common of all are government-managed forestry projects: fuelwood projects; watershed programs; and national schemes for reafforestation that may feature erosion control, production of wood for construction, growing of fruits or other tree-borne food products—and often only incidentally the enhancement of fuelwood supplies. (Some government projects are more in the line of parastatal business entreprises and are treated above in chapter 10.) Often international donor agencies are directly or indirectly the managers of bioenergy-related projects. Governments and donors have also shown a growing interest in promoting the formation of associations of farmers to produce biomass feedstocks for energy or fiber.

Other programs have been established, again usually under government auspices, directly aiding smallholders such as small farmers to produce bioenergy. For example, small farmers could grow wood for sale to industries; we have already seen in chapter 10 how such smallholders were included in a scheme developed by commercial steel mills in Brazil. A good deal of effort in forestry in general, and often specifically in fuelwood bioenergy, has also gone into encouraging traditional farmers to grow trees for household use on unused land or in combination with field crops on arable farmland. Not only farmers, but institutions such as schools and churches have often undertaken bioenergy efforts. Other bioenergy projects have been promoted at the household consumption level, encouraging the installation of household biogas plants or the use of more efficient stoves.

Government-Managed (Or Donor-Directed) Projects

A vast number of bioenergy programs, or programs touching on bioenergy, are not under the management of individuals or even communities, but are under the direction of "outsiders." These outsiders may be agencies of the national or regional government, or they may be international donors acting with the cooperation, or at least enjoying the tolerance of the government agencies concerned. Often the government itself is the logical entrepreneur because, as in many areas of economic development, if something is not done by the government, it is not done at all. Often only the government or an outside "quasi-government" such as a foundation or other foreign assistance outlet will be willing to undertake a project because certain benefits from the project will not accrue to any private individual. National afforestation programs, for example, can involve regional or community goals by promoting "protective afforestation"—that is, growing trees as protection from water erosion, planting shelterbelts against wind erosion, and sometimes cultivating species that provide shade or browse for grazing animals. In addition, generating rural jobs will often be cited as the justification for forestry projects in areas with much underemployment while normal kinds of forestry benefits like providing poles and fuelwood may appear to be almost afterthoughts (Evans, 1982, p. 42). Experts have even criticized international donors—not often accused of Scrooge-like behavior—for restricting project analysis criteria to economic viability (FPL, 1980, pp. 6-7), slighting less quantifiable benefits such as control of soil erosion, biomass for grazing, and the providing of habitat for harvestable wild animals.

The use of the government or an outside donor as the entrepreneur involves what is sometimes a fatal anomaly. The government is often the only one that can do the job at the moment, because it may be the only one with the resources, foresight, or interest in multiple objectives to carry out particular projects. Yet, except for the centrally-planned economies, the government efforts are more often than not thought of as only "pilot" or "demonstration" projects. The explicit ideal is generally that such projects will eventually become self-sustaining, and that private industry, small farmers, or associations will eventually take over the project technology and see to its wider dissemination. This "hand-over" supposition tends to be even more entrenched when the projects are sponsored by an international donor, an entity that will often—though by no means always—be involved only on a temporary basis in one specific project in one particular country. Therefore in our review here of different methods of organization, it is not always possible to identify with certainty a project as "government-managed"—because the government management may be envisaged as a temporary expedient for one demonstration plant or as a makeshift substitute for some other, final organizational set-up. Furthermore, as we have seen in both chapters 10 and 11, governments in developing countries by custom play large roles in areas like agriculture and forestry that touch on bioenergy. And as we shall see below, many projects that are at least formally managed by associations, smallholders, or others will have intimate connections with government agencies.

In chapter 10 we considered large government bioenergy operations run as enterprises. Here we examine a few examples of other types of government-managed projects, usually more oriented toward development than business, pointing out some features of interest to bioenergy planners organizing or evaluating future Third World projects. For example, desert reclamation belongs to the normal type of activities that are customarily carried out by governments. The Peruvian government has set up pilot projects planting *Prosopis* (mesquite) in sandy soils in order to stabilize the soil and make agriculture possible (FPL, 1980, p. 76). Resettled families are given two hectares for growing crops like pigeon peas and watermelons. They are also participants in a bioenergy experiment, because they get fuelwood from the mesquite.

A World Bank project in Bangladesh displays some of the complexities typical of ambitious officially-sponsored projects. The project contained, for example, a land use dimension: mangroves were planted as a protective barrier against wind and waves to accelerate land accretion in the coastal areas. There was an equity dimension: by creating more land for agriculture, the project created new farms that could be deeded to landless tenant farmers (World Bank, 1982I, pp. 42-44). It also produced timber for industrial wood markets, and the planting provided fuelwood, which tended to replace kerosene as a household fuel and therefore helped conserve foreign exchange.

Such governmental programs need not be confined to the production of wood. We have seen in chapter 11 how community biogas programs in India are for the most part not quite "community," but are predominantly implemented by government agencies. Furthermore, some conventional government forestry projects in support of agriculture such as woodlots planted primarily as shelterbelts in the northern part of the Sudan could also supply modern biofuels. Some auxiliary plots in the Sudan projects , for example, are planted primarily for wood production and could provide a feedstock, either woodchips or charcoal, for gasifiers to produce a badly-needed replacement fuel for diesel-operated irrigation pumps (Sudan, 1982, p. 9).

In a number of cases, governments have sought to provide employment to rural workers as part of reforestation efforts; the COPLAMAR project in Mexico was one of this type, paying local laborers for planting and other activities (Thrupp, 1981, p. 4). In a Forestry Department project in Mexico, the government also hired members of a "Civil Forestry Group" at two to three pesos per tree planted. The project utilized land that was private, communal, or state-owned; its scheme for dividing the harvest was one-third of the harvest to the owner, one-third to the workers, and one-third to the community (Thrupp, 1981, p. 3).

In some cases, although rural inhabitants were put on government-financed salaries, the motivation was not seen as unemployment relief but as an "advance" to tide the farmers over the years before the revenues from forestry materialized. For example, in the Instituto de Recursos Naturales Renovables (RENARE) project in Panama, each farmer was given a salary plus two hectares of state-owned land on which he was to do subsistence farming while also planting trees (Thrupp, 1981, p. 25). In some cases, the problem of delayed returns was attacked by encouraging villagers to grow trees instead of food crops and by paying them for work on the tree farms with U.S. surplus food (Thrupp, 1981, p. 41). Under this scheme, the employment and payment in kind aspect is not seen as a permanent feature: the Peruvian government will own the trees for only 5 years and will control usage through permits for another 10, but then the ownership will pass to the landowner or to the community.

Unfortunately, there are often problems both in planning and execution. In the COPLAMAR case in Mexico, there was a notable deficiency in planning for the eventual end use of the wood. In the RENARE case, the inputs of help from the RENARE staff turned out to be very heavy, and over-dependence was identified as a possible problem. In the Peruvian case, although the food incentive is to terminate eventually, there was apparently little idea of how a continuation of tree maintenance could be motivated.

In Morocco, government-run schemes were developed to produce feedstock for a pulp mill while providing income to nomadic families (FPL, 1980, pp. 64-65). Although the wood was harvested by salaried workers, it was sold to contractors, who were obligated to resell 70 percent of the wood to the pulp mill but could resell the other 30 percent at market prices for use as poles and fuelwood.[1]

Often, if a government happens to develop a positive and creative attitude toward the question of bioenergy—in most past cases, forestry—it may become difficult to tell where government operations end and private or community operations begin. The massive reforestation program in Korea is a good example. In addition to the community-oriented Korean programs discussed in chapter 11, the government also played a key role in the promotion of tree planting on private land. Under the laws establishing the program, the government could either have the local Village Forestry Association farm private land on an arbitrary though compensated basis, or it could appoint an "executor" to manage tree farming there (FAO, 1982, p. 29).

In the state of Gujarat in India, the aggressive program of the state forestry department has taken many forms, as seen in chapter 11 and also below. In addition to the programs involving communities and individuals that the forestry department both promotes and aids with technical assistance, it also operates its own tree plantations. In particular, major stretches of roadsides in Gujarat along the primary, secondary, and tertiary road systems of the state have now been planted with trees of varying species suitable for fuelwood. In addition to the roadside plantings, there are various plots of unused land behind the roadways that have been planted by the Forestry Department with eucalyptus to produce poles and fuelwood (Java, 1984a). One of the advantages of this program, operated solely by the government with official staffing, is that new ideas can be tried—current experiments utilizing 20 fertilizings and 20 irrigations to produce harvestable eucalyptus on a three-year rotation, for example.

This is not to say that the Gujarat Forestry Department efforts are run entirely independently of the surrounding communities; the cooperation of the local inhabitants is necessary, for example, to discourage poaching. Also local help is needed in preventing sapling destruction by grazing animals— although the use of species of spiny succulents (mostly *Euphorbia tereticornis* hybrids) as "natural fences" helps protect the young trees. As part of its effort to gain community cooperation, the state government had long planned to split the fuelwood production from roadside trees on a 50-50 basis with the local village councils (*panchayats*) (Java, 1984, p. 6). Future policy, however, may opt instead

1. The program is not without its critics. Complaints have been recorded about the shortage of qualified foresters and the consequent lack of progress in raising yields, and the low level of the wages paid as a result of the low fixed price paid by the mill for pulpwood. The eucalyptus species used had also inhibited drainage in some cases, and the toxic chemicals it emits have inhibited the growth of browse for livestock.

for using the wood harvests to help alleviate rural poverty by doling out wood on ration cards, just as is done with subsistence rations of rice, kerosene, and other important staples in India (Java, 1984a).

The pitfalls of government management—lack of continuity and motivation—are well known. The government route, however, can be viable if planning is comprehensive and if creative steps are taken toward community involvement.

Associations

In a different mode of attack on the organizational problem, groups of developing area farmers or smallholders band together in various associations or cooperatives dedicated to the cultivation of trees for bioenergy or other purposes. In general, such associations will have at least tacit government approval—or may often, in fact, be creatures of government policy. The distinction between an "association"—a more or less voluntary grouping—and a "community"—the collectivity of a locality—can in fact be sometimes so fine as to be meaningless. Indeed, the resemblance between the Philippines "dendrothermal" program (see chapter 10 and below) and the Korean program with its Village Forestry Associations is quite close, the associations being in effect community organizations.

The dendrothermal program in the Philippines displays a high degree of complexity in the relationship between government agencies and the rural "associations" they sponsor. The two most powerful actors in the dendrothermal policy arena, for example, have been a government and a quasi-governmental entity: the National Energy Agency (NEA), in charge of general energy development, and the Rural Electric Cooperatives, or local electric utilities. The local Tree Farmers' Associations, however, are a key element of the scheme and are charged with the management of the feedstock production. The associations will also share the management of the power plant operation, along with the NEA and the Rural Electric Cooperatives.

The key to the viability of the associations is, however, motivation. A study of the farmer associations has shown that poverty was, unsurprisingly, one of the best motivations for joining the project. Regardless of income level, however, most participants seemed to have a strong motivation toward change and self-improvement (Denton, 1983, pp. 67, 80, 82).

Another innovative program in the Philippines using local associations is oriented toward supplying energy for agriculture. A new wood gasification project seeks to replace imported diesel fuel now used for irrigation pumping. The funding for the project will be supplied by the Farm Systems Development Corporation (FSDC), an autonomous public corporation. The FSDC organizes Integrated Service Associations (ISA) of about 60 farmers each (AID, 1982, Annex II, p. 1). The ISA members receive loans from the FSDC to buy equipment for converting the diesel-operated pumps to fueling from wood gasifiers (AID, 1982, Annex II, p. 4). The ISAs, moreover, have the main administrative function in organizing the wood feedstock supply. They lease land—usually in six-acre lots of marginal quality—and then they recruit families to grow woodlots on these lands and perform wood chipping operations (AID, 1982, Annex II, p. 14). These families are mostly landless laborers and are paid wages by the ISAs. The FSDC also supplies loans for funding the start-ups of the woodlots.

As part of the very active program of tree growing in the Indian state of Gujarat, the forest department there is cooperating with a growing number of "forest societies," spontaneously formed voluntary membership groups that are growing fuelwood and pole trees commercially (Java, 1984, p. 7). The societies will cultivate trees alternately in rows with their field crops. They are to take advantage of one of the main virtues of the cooperative system—as seen in the case of the cooperative formed by the Hawaiian sugar growers in chapter 10—the enhanced availability of finance. The State has induced the nationalized banks to advance the necessary loans of 5 thousand rupees ($500) each, at a low rate of interest, to each of the approximately 4 thousand members of the local societies. If the project works out, one of its problems could turn out to be too much success: there is the possibility of an over-production of poles for construction purposes. In fact, the activities of local milk cooperatives in avoiding market gluts in the field of milk sales is currently being studied, with some interest in the eventual development of similar mechanisms to support wood prices (Java, 1984a).

Many of these association schemes are somewhat new, and it is difficult to determine their prospects. The types of organization considered represent some creative thinking in promoting bioenergy, however.

Smallholders: Individual Farmers and Entrepreneurs

There have been numerous efforts to help individual farmers develop bioenergy resources without resorting to formal organizational structures. In fact, a recent fuelwood project in Mali emphasized individual plots near family compounds because of past failures of communal schemes (World Bank, 1980d, p. 18). Furthermore, farmers are often spontaneously involved in tree growing, as after the beginnings of a local fuelwood shortage, small hedgerow or shade plot plantings were observed to spring up in the Embu district of Kenya (Brokensha and Riley, 1978, p. 26).

In some cases farmers have been encouraged to carry out bioenergy-related activities as an adjunct to agricultural activities. In other cases, there has been a major redirection of effort and capital by the farmer to bioenergy. Sometimes bioenergy development has been included as part of a larger farm-based tree program for fiber or fruit production. For example, in Rwanda there has been an emphasis on improving forestry extension to encourage the setting up of rural woodlots to be owned by individuals (World Bank, 1982y). In Costa Rica a government-sponsored demonstration project divided the labor requirement between the government and the smallholder (Thrupp, 1981, p. 17). In this scheme, the local forestry service provided technical assistance and 50 percent of the labor, as well as supplying and maintaining the seedlings. In Indonesia, government incentives encouraged farmers to plant *Calliandra* for fuelwood, trying to make the farmers both self-sufficient in wood and encouraging them to grow surpluses to meet the needs of local industry (FPL, 1980, p. 60).

The smallholder option has sometimes experienced difficulties. In one much-touted scheme in the province of Ilocos Norte in the Philippines, the net result was that 89 percent of the barangays (communities) had planted less than three hectares in trees (Hyman, 1982a, p. 18). The smallholder option was rejected in a case in Pakistan because the small size of family land-holdings made action by the forest department impractical (Cernea, 1980, p. 26).

The role of competent and aggressive help on the part of forestry extension services is exceedingly important in preventing failure of smallholder schemes, if the experience in Gujarat state in India is any example. The Forestry Department there has promoted strip planting by farmers alongside fields by providing seedlings and comprehensive technical assistance (Java, 1983, p. 6; Ranganathan, 1981, p. 11). Individual local farmers have in fact planted as many as 8-10 thousand eucalyptus on marginal lands on the edges of their cropland (Java, 1984, p. 8). The size of the program is large, with approximately 100 thousand farmers involved and with some 250 million seedlings having been given out within the past few years. As noted in chapter 10, a few farmers have even gone into the wood-growing business professionally. Various expedients have been created to solve particular problems. For example, one problem is getting the seedlings to the farmers in a timely manner before the rains start: in recent years, schoolboys have been hired to deliver the seedlings for a fee of 20 paise (two cents) apiece (Java, 1984a).

Traditionally, a great number of smallholder cottage industries in developing countries have operated their own woodlots to provide fuel for tobacco drying, sugar manufacture, or pottery kilns. Efforts to help such industries by encouraging tree farming for small-scale tobacco curing, for example, were reportedly not successful in the Philippines for many of the reasons that customarily plague such projects: local cultural reluctance to borrow, inadequate forestry extension services, labor management problems, etc.—plus more exotic difficulties such as arson motivated by local jealousies (Hyman, 1982, p. 36).

Results have been more promising for other smallholder projects that—perhaps not coincidentally—involved elements other than fuelwood. In northern Sudan, for example, fuelwood production has always been a small but important part of a fairly complex traditional system of agroforestry. In this system, various field crops are rotated on the sandy soil of a semi-arid region, and the crop fields are bordered by trees of the species *Acacia senegal*, which produces the valuable chemical commodity gum arabic in addition to serving as a soil stabilizer and source of fuelwood (Hammer, 1982, p. 4). The traditional system depended on letting the land lie fallow several years and then, toward the end of the fallow period, harvesting the sap from the trees for gum arabic. But growing desertification—owing ultimately to weather fluctuations or climatic changes and overpopulation—has led to a decrease in fallow periods, excessive grazing, and other secondary response factors that have damaged soils and reduced harvests of both tree crops and field crops (Hammer, 1982, pp. 6-9). So new projects to replant *Acacia senegal* have been brought forward within the last few years. These projects have benefitted from having a well-understood economic motivation: the reestablishment of a familiar, specialized, and profitable agricultural business. A total of 973 farmers—compared to a target of 240—planted seeds and seedlings during the first rainy season (Hammer, 1982, p. 52).[2]

Organization made a difference in the Sudan. In villages where there were well-organized community societies serving as adjuncts to the project, seeds were shared more efficiently and equitably between interested farmers. In other villages with little sense of community, the few most active and powerful farmers tended to be favored and received a disproportionate share of project benefits (Hammer, 1982, p. 52).

2. Water and fuel, however, remained the major basic needs concerns of the villagers: this fact proved to be a great handicap to the project. The study quoted suggests that an integrated attack on the village problem—an approach that was indeed first proposed during the *Acacia* project—would be the best approach for new development there (Hammer, 1982, pp. 63-64).

Participation by women in the gum arabic development projects, although opposed by many traditional male interest groups, also turned out to be a crucial factor. Women not only fulfilled an important role in helping their husbands farm, but many farms were owned by women—especially widows. In fact, the village committees set up especially to help coordinate the project included 3 to 5 women on their rosters of 7 to 15 persons (Hammer, 1982, p. 44).

The successful Paper Industries Corporation of the Philippines (PICOP) project was primarily designed to provide pulpwood for domestic production of newsprint (Hyman, 1982, pp. 2-3); however, the project included a fuelwood component.[3] The goal of the project was to grow trees on marginal land, with 20 percent of the funds (loans) to be spent for interim crops or livestock as part of an agroforestry component. Some casual labor was hired and the harvesting was usually contracted out (Hyman, 1982, p. 7), but all of the growers of trees were landowners, and the households provided most of the labor. The success of the project depended on some special factors and some of more general interest.[4] One prominent condition for success was the existence of an assured market (Hyman, 1982, p. 31). Also, technical assistance and supervision were effective and seedling mortality, for example, was kept low. Finally, the project was perceived locally as helping both rural development and the environment. There were also some especially helpful pre-conditions. Some of the participants in the PICOP project had done similar projects, and there was already an existing infrastructure available to provide seedling supply and tree growing expertise.[5]

Some useful lessons could be inferred about credit from the PICOP experience. For one thing, it appeared that loans should have included modifications to reflect inflation so that project funding would not fall short. For another, harvesting costs turned out to be unexpectedly large and additional loan funds might well have been set aside for financing harvesting (Hyman, 1982, p. 23).[6]

Institutions

Institutions such as churches, schools, and hospitals play an important if sometimes little recognized role in the development of new

3. PICOP operated the paper mill and had tried independently to stimulate local wood production; World Bank support made the project possible.

4. It must be noted, however, that the project did not go quite as planned. The trees were not planted in strict rotation as had been the design; in fact the financing banks even *required* immediate clearing of all land in most cases. In addition, the agroforestry approach was dropped because the original loans were insufficient to provide the necessary funds; about 7 percent of the farmers, however, intercropped on their own lands (Hyman, 1982a, pp. 8-10).

5. Note that the project may not be judged entirely successful on equity grounds, in that the average household involved had an income of $8,700, while the 1980 average income in the Philippines was $3,200 (Hyman, 1982, p. 6).

6. Another special circumstance was that the pulpwood price was judged by some to be too low.

energy technologies, including bioenergy, in the Third World. As part of the comprehensive reafforestation program in the state of Gujarat in India, for example, one college was encouraged to plant as many as 2 thousand trees on its campus. Even at 45 rupees ($4.50) a tree, receipts from sales of wood for fuelwood and poles, after subtracting approximately 10 percent for the cost of tending and planting, enabled the college to turn a tidy profit.

Institutions can be exceedingly useful in demonstration projects both because of their social visibility and because they can sometimes carry out projects on a scale larger than the ordinary smallholder. In addition, institutions are often amenable to direction from central agencies and can therefore be easily steered into new bioenergy fields. For example, in India, as part of the KVIC biogas program, 25 military installations have been ordered to install biogas facilities at their dairy farms; four of the dairy farm digesters are already in operation (Chopra, 1984). In one area in Tanzania, the Anglican Church took up the cause of tree planting, opened a small nursery, and distributed seedlings to churches. Seedlings were also sold to individuals and, in fact, were sometimes stolen at night—a not entirely undesirable outcome under the circumstances (Skutsch, 1983, pp. 36-37).

Household Biofuel Production: Biogas

One of the most prominent uses of new types of bioenergy conversion facilities at the household level is the biogas digester. In India, plants number as many as 100 thousand, with 30 thousand in Korea, and significant numbers in such countries as Taiwan, Nepal, the Philippines, Indonesia, Thailand, and Pakistan. A number of biogas projects are being financed by AID and other agencies on all continents of the developing world (Mahin, 1982, p. 2).[7] The oldest program is that of the Khadi and Village Industries Commission (KVIC) in Bombay and is quite comprehensive in scope. The KVIC staff will help farmers— those owning at least two head of cattle[8]—obtain loans from the nationalized banks. They will also give technical guidance on the installation of the plant and arrange for the supply of appliances and of accessories (KVIC, 1983a, p. 3).[9]

7. Biogas units have also been widely used in connection with "integrated farming schemes." As mentioned in chapter 10, projects have been implemented in Fiji, using designs combining livestock, aquaculture (with algae), and ordinary cropping. They used animal wastes as a biogas plant feedstock and its effluent as an aquaculture nutrient source, while the aquaculture effluent was used as water for field crops (Chan, 1981, pp. 126-127).

Digester slurry has also been used to grow high-protein algae in the Philippines and Israel (Mahin, 1982, p. 21).

8. Biogas for households has been widely criticized because only relatively well-to-do households will own enough animals (three to five) to produce an adequate supply of gas. This intellectual burden of redressing equity problems should however properly be shouldered by all noncollective options in poor subsistence economies.

9. The KVIC also runs a "self-employment scheme" in conjunction with the biogas (Gobar Gas) program under which unemployed persons learn to construct gas plants at first on an apprentice basis and then as a "certified trained worker."

Biogas technology has been introduced in Thailand under the aegis of the Thai Department of Health. Despite publicity and promotional efforts, however, the technology has not been widely accepted by rural people, and less than 50 percent of the biogas plants that have been installed were being used, according to one evaluation (Ratasuk, 1981, p. 1150). Difficulties with the digesters and a shortage of feedstock were common technical problems that were probably exacerbated by inadequate liaison.

Similarly, a review of work in the Philippines suggested that biogas systems, despite their "simple" nature, required skilled attention and maintenance (Terrado, 1981, p. 161). The ordinary homeowner had neither the background nor the patience to troubleshoot units. Even official demonstration plants like those run by the Bureau of Animal Industry, with full-time caretakers, suffered many plant shutdowns due to technical problems. The possibility of alleviating this problem through aggressive extension work was considered, but the general unpromising economics of household units in the area apparently discouraged the government from working on improvements in institutional organization.

The organizational problems involved in spreading such a technology are evidently not restricted to free market economies. The total number of digesters built in China has been quite large (over 7 million as of mid-1979), and training methods of the "each-one-teach-one" variety have evidently been effective (Taylor, 1981, pp. 191, 197). Nevertheless, subjective factors like the attitude of local leaders have evidently played a key role in whether or not biogas digesters succeed in a given locality (Taylor, 1981, pp. 195-196).

Household Consumption: Stoves

Perhaps one of the most important "sources" of biomass energy could be the conservation of energy from wood, charcoal, or wastes by the introduction of newer and more efficient stoves.

In a campaign carried on by the Choqui organization to spread use of the *Lorena* type of improved wood stove in Guatemala, promoters succeeded in installing more than 300 units. They also attempted to spread the technology by sponsoring short courses in stove construction; they estimated that 400-500 stoves had been built as a result of the training (ROCAP, 1979, pp. 99-100). Project monitoring, however, was evidently haphazard, while the presence of widespread technical defects evidently failed to stimulate project redesign.[10] The agents of

10. During an evaluation tour of 34 stoves installed by the *Choqui* technicians, it was discovered that 73 percent of the stoves were cracked, 58 percent leaked smoke, and 9 percent had become inoperable (ROCAP, 1979, p. 400).

the program claimed that 95 percent of the families questioned said that they were very satisfied with their stoves; on the other hand, there was reportedly little general acceptance of the stoves by the general population.

In contrast, through a combined effort between Centro Mesomerico de Estudios Sobre Tecnologia Apropriado (CEMAT) and Appropriate Technologies International (ATI), 400-500 *Lorena* stoves were to be built in Guatemala; a follow-up survey found that of 58 families that reported they used the stoves daily, 29 families, or 57 percent, reported fuelwood savings of 50 percent or better (Morgan and Icerman, 1981, p. 101). Another project sponsored by the Center of Appropriate Technology in Nicaragua carried out demonstrations in rural areas by working through existing rural associations (Thrupp, 1981, pp. 12-13). The demonstrations were rated as successful, and the women were reported happy with the *Lorena* stoves.

An improved *Ondol* (pit) stove design was developed by the Forest Research Institute (FRI) of Korea (FAO, 1982a, pp. 74, 91). Initial resistance to the modification was reportedly overcome through a training program in which hundreds of extension workers used literature and demonstration materials supplied by the FRI to explain the improved *Ondol* to householders. By 1979, some 90 percent of existing stoves were said to have been upgraded.

Increasingly, interest in stove improvement centers on small, portable stoves made of ceramics or metal, especially for use with charcoal. These can be manufactured by local artisans and marketed through normal channels. The "Thai bucket" and similar derivatives have sought to combine the lightness and structural strength of metal with an insulating layer of ceramics (Foley and Moss, 1983, pp. 91-92, 113). For example, Kenyan versions have gone into production on a fairly large scale and have even suffered from quality-control problems as local makers sometimes radically modify designs to cut costs.

Stove technology is an area where gender problems in development are apt to surface. The private "Women for Progress" group in Kenya was assigned, for example, a formal role in a government program for improving charcoal and wood stoves (Thrupp, 1983, pp. 28-29). Nevertheless, it was reported that male planners in the Ministry of Energy, although concerned about women's preferences in the use of new stoves, rejected suggestions that women's groups should be involved in choosing future stove technologies. The consequences of such attitudes in this particular case are not yet clear. But the example may serve to remind the bioenergy planner of the key role of consumer—including gender—preferences in any successful household fuel conservation project (See Appendix 11-A).

Miscellaneous Organizational Approaches: Conclusion

A key aspect of this chapter on miscellaneous organizational approaches is government intervention in the promotion of biomass energy. Sometimes governments have acted directly, as employers of workers for bioenergy projects. Sometimes they have acted indirectly through associations or through providing aid for action by individuals, associations, and institutions. But there has also been much work done independently of governments, especially by institutions. In fact, the success of programs initiated or aided by government efforts probably depends on obtaining the same type of cooperation and local interest that is shown in certain private and institutional efforts. Although local circumstances vary—see the complex economic and social situation in the *Acacia senegal* project in the Sudan—it is evidently important to design consistent and long-term programs with proper attention to feedback from bioenergy producers and users.

Planning Procedure: Comparison of Organizational Alternatives

In the last three chapters, we have examined the commercial and community approaches and other approaches to organizing bioenergy projects. How can the relative advantages of these approaches be judged so that the best choice can be made for plausible bioenergy options?

One of the most straightforward approaches is to encourage the growth of commercial bioenergy firms. A great advantage is that ideally the operation runs itself. The difficulty is that with some exceptions the private sector has not seen most modern bioenergy applications as economically viable. Of course, that is not necessarily an argument for choosing a different type of organizational setup, because economic viability will be important in community and in smallholder schemes too. At any rate, national policy goals could justify subsidization of commercial biofuel production.

Bioenergy options that can be carried out economically as byproducts of an ongoing commercial operation also have the advantage of using built-in entrepreneurial motivation and skills. On the other hand, if operations are really economically feasible, there is less need in the first place for project interventioin from the outside. Furthermore, not all byproduct operations are successful, as has been seen with some biogas schemes.

One of the advantages for the planner in using community management is that unfavorable economic situations may be overcome—or at least be compensated for. If hidden local resources such as underutilized labor are available, results of the community approach should in principle be good. Furthermore, building community spirit for other development purposes and achieving greater income equity are worthwhile goals that could justify some economic penalty. The great disadvantage tends to be that a spirit of community cooperation, rather than the result of a project, may often constitute a prerequisite for it.

Other modes of management, by the government or by individuals, may suffer from the problem that management is often not sufficiently capable or lacks motivation or both. Some options may be able to get around this problem by enlisting the aid of institutions like schools and churches. In addition, newer tendencies toward using associations of farmers or entrepreneurs could form some kind of middle ground between the individual and commercial approach.

13

Bioenergy and National Planning

Introduction: Planning and the Overlap Problem

Bioenergy is an area in which there is overlapping government responsiblity: it deals with issues relating to such diverse fields as agriculture, environment, trade, finance, and land tenure. Complex long-range decisions may be involved. Before planting forests, for example, a country may have to train forestry experts. Only planners at the national level have the legal authority, financial resources, technical support, and political and management acumen to develop an efficient and comprehensive energy plan. The bioenergy project planner must be able to interact efficiently with such national-level efforts. The problems of bioenergy relating to *overall* national planning are treated in this chapter. The relation of bioenergy projects to national *bioenergy* policies is discussed in chapter 14.

National planning is necessary to resolve the basic contradictions in bioenergy development, e.g. short-term imperatives versus long-term costs, food versus fuel trade-offs, individual needs versus community interests, and urban versus rural imperatives (Hall et al., 1982, p. 147). An answer to a problem in one of these areas may result in an added cost in another if an integrated approach is not taken. If vegetable oil is used for fuel, it will not be available for cooking and thus will affect food markets and might affect nutrition levels.

> Comparable in complexity to the management systems associated with "green revolution" agriculture, family planning, or rural health care, bioenergy system management will reflect not only economic and site specific features but also social, cultural, and political considerations as well as the structure of the government organization responsible for the bioenergy program (Weatherly, 1983, p. 2).

A program must relate its major components to national priorities at the same time it presents meaningful investment options to lending agencies and defines the role of participants in accepting risk (Weatherly, 1983, p. 3).

As a primary goal, the planner must assess which policies are most beneficial to the society as a whole. The trade-offs between different policies must therefore be examined. A typical trade-off relevant here is between trade promotion—e.g., expansion of export food grains—and energy substitution through bioenergy production. Which goal, or mix of goals, does the society want to pursue? The answer to this question requires an investigation of the linkages between bioenergy and the rest of the economy.

In this chapter we will assess the connection between bioenergy and national planning in three broad categories: economic, environmental, and social. The key economic question is how dedication of scarce development resources to bioenergy adversely affects other national objectives. That is, what are the various trade-offs in terms of food, fodder, fuel, and fiber. The most severe environmental issues concern the impact of bioenergy development on soil and air quality. How will these environmental impacts in turn be felt as impacts on health, agriculture, and water supply? An analysis of the various social impacts of bioenergy development is also necessary. How, for example, will land tenure affect or be affected by bioenergy development?

Economic Trade-offs at the National Level

At the national level the economic tradeoffs associated with bioenergy development can be assessed in terms of three main issues: foreign exchange constraints, competition for factor supplies, and the impact of pricing policies. The net foreign exchange gains (or losses) from bioenergy schemes, the effect of bioenergy development on scarce resource allocation, and the likely impact of product pricing policies for biofuels and its competitors on factor allocations may be crucial decision variables for the planner.

Foreign Exchange Trade-offs

A major motivation for developing indigenous energy resources is to save scarce foreign exchange. A bioenergy project developed to substitute for a declining local energy resource, such as fuelwood, or to substitute for imported oil has the same rationale in both cases: a failure to develop bioenergy will ultimately lead to an increase of expenditures for imported energy. But there are alternatives to bioenergy that also save foreign exchange. A long-run strategy to expand exports could in some cases offer a more efficient method of improving a nation's balance of payments situation than a diversion of resources to oil import substitutes.

Bioenergy can then have a direct impact on a country's balance-of-payments picture by substituting for imported oil. Liquid biofuels, such as alcohol or oil producing crops, can substitute directly for petroleum. Bioenergy can also substitute for oil indirectly through the replacement of oil by wood in electricity generation and by charcoal in industrial heat (Hall et al., 1982, pp. 120-126). But this gain can become a loss under certain conditions. If the shift in production adversely affects another foreign exchange earner or requires funds to be expended on other imports, the flow of foreign exchange may be reversed. For example, the use of sugarcane for alcohol production will eliminate a portion of sugar exports. Also, dependency on food imports may be substituted for an energy import dependency if land for food production is used for fuel production. In addition, new capital imports may be required for machinery to produce the fuel. For example, in Brazil a policy question of great interest is which product helps most directly in improving the nation's balance of payments: sugar or alcohol? One analysis of Brazil's PROALCOOL program concluded that the long-run potential of sugar as a foreign exchange earner was negligible in comparison to potential foreign exchange gains in substituting alcohol for imported oil—even though a foreign exchange loss was incurred in the short run. Such a short-run loss could conceivably be considered an acceptable additional penalty or opportunity cost on alcohol production (Islam and Ramsay, 1982, p. 21).

Although this last conclusion could be contested even in the Brazilian case, it does illustrate the necessity for a dynamic analysis in assessing the foreign exchange balance in terms of long-run demand patterns for export crops. The resource trade-offs necessary to accomplish export promotion or energy substitution programs constitute necessary but not sufficient information; the structure of the export market must also be understood (Islam and Ramsay, 1982, p. 2).

Production Factor Competition

How can a country meet its export crop and energy production targets in the most cost-effective way given basic constraints such as land and capital? Industrial forest products (timber, newsprint, packaging, and paperboard) are, for example, important to world trade. Will a bioenergy project compete with them for scarce inputs? If land is the key constraint, how long will it take to bring new land of varying degrees of productivity under the plow? Existing constraints on productivity in other sectors will play a key role. Is the cement industry, for example, held back by high prices for energy or for other factors such as capital, markets, or technical and managerial skills (Wionczek et al., 1982, p. 11)? There are thus two main areas of interest: constraints on trade promotion and the impact of diverting resources from trade promotion to energy production. For bioenergy, the trade-off is often direct, that is, between producing nonenergy agricultural goods and diverting the same inputs into agricultural production of bioenergy.

In order to answer these questions definitively, data would be needed on bioenergy production costs, how much alternatives cost, and whether costs and other constraints render the available alternatives attractive or unattractive to producers (Lockeretz, 1982, p. 3). The element that links bioenergy to the rest of the economy is price. The prices (costs) of economic factors determine the level of factor substitution—e.g., how much labor goes into which energy or nonenergy sector at what level of wages (labor price). The situation will change over time, especially as the agricultural sector is required to be both an energy consumer and producer. Furthermore, it depends on efficiencies and crop yields, because at least one study has found that low agricultural productivity tends to be more energy efficient than high agricultural productivity (Green, 1978, p. 8).

The basic resource trade-off in bioenergy is often seen as the commitment of a scarce resource (usually land) to fuel production at the expense of food production. But the question is not simply a conflict of food versus fuel but a competition between all alternative resource uses: food, fuel, fodder, and fiber. For example, forestry projects to promote fuelwood production may displace not only animal and human food production but also another source of fuel and fertilizer: crop and animal wastes. Crop residues are used for fuel as well as for animal feed and roof thatching. Animal dung is a source of fuel as well as a building material; and of course the diversion of dung to use as fuel will affect food production in many countries unless a substitute soil nutrient is found. The FAO estimates that dung burning already eliminates 20 million tons of food grain from world production each year (Spears, 1980, p. 5).

Land. The most obvious production constraint encountered is land availability. To have a major impact on reducing petroleum imports, a bioenergy program needs significant supplies of unused or underutilized land (Hall et al., 1982, p. 119). If land for food production is directly substituted for fuel production, food prices will tend to increase. Withdrawing either food or food-producing land from food and feed markets will place an upward pressure on the price of land, the factor whose supply has been reduced (Lockeretz, 1982, p. 53). In Brazil it was discovered that an expansion of sugarcane-producing acreage for alcohol production does indeed affect demand for land. Unless this competition occurs in a region characterized by a large inventory of "frontier" land, land prices will increase (Pelin, 1980, pp. 1, 4). As a result, it has been argued (Pelin, 1980, p. 25) that the Brazilian program will produce positive benefits only if it utilizes mainly frontier lands.

Differences between types of land therefore matter; for example, a useful difference can be drawn between frontier and marginal lands. Frontier areas are fertile lands that, primarily because of a lack of population pressure, have not been developed. They are "transport-limited" and development may hinge on new road networks; as we already discussed in chapter 9, new roads may be built without any assessed cost to the frontier lands. Marginal land can be identified as those having low productivity under present inputs of other on-site factors. Development of marginal lands will require greater inputs—such as fertilizer—thereby directly and inevitably raising production costs.

Even though land may not be a limiting factor in one region of the country, it may well be in another. In Brazil, sugarcane production displaces food production at a municipal level because municipalities depend greatly on locally produced food and the local food distribution system is not set up for long-range transportation (Poole, 1983, p. 6). As we will see below, in India the concentration of a sugar-based alcohol fuels program in one region resulted in significant food price increases (Bhatia, 1981, p. 33).

The project planner may therefore need to determine what crops an energy feedstock competes with for the use of fertile land in order to determine the seriousness of potential bioenergy impacts. Under a promotional policy in India, medium-sized farms expanded their cane acreage at the expense of food grains for human consumption while large farmers substituted sugarcane for fodder crops (Bhatia, 1981, p. 27). In some countries or regions only one crop may be displaced, while in others with a longer growing season it may be more than one.[1] Different crops have different costs of cultivation and there are even differences within crops, such as planted versus ratoon (resprouted) sugarcane. A sugar crop harvest cycle varies from 12 to 20 months (Bhatia, 1981, p. 9), and different farmers may, for this and other

1. Labor needs may also be increased. In one Indian district, sugarcane requires two labor units for every one used in the production of *either* barley *or* wheat. Thus if sugarcane replaces one traditional crop, there is a net gain in employment; if it replaces three or more, there is a net loss (Bhatia, 1981, p. 30). A Brazilian study also found that alcohol production could be more competitive in labor surplus areas (Rask and Adams, 1979, pp. 10-11).

reasons, take different amounts of time to readjust their acreage allocations between crops to new price changes.

Other Factors. What if land as a physical resource is not the scarcest production factor? Is there a zero opportunity cost for land for bioenergy—that is, can production take place without a significant loss of alternative revenues—as has been suggested for Northeast Brazil (Silva, 1982, p. 60)? No, because although land may not be a scarce input, other factors may be, including other materials such as water and fertilizer, and personnel such as agricultural technicians. Indeed, some see availability of fertile land as not the key problem in Brazil (Poole, 1983, p. 34). Instead, a reluctance on sociological ("cattleman-sheepherder") grounds to change land use from pasture to farming and the limited national supply of farmers and other entrepreneurs could be the major reasons for future pressure on agricultural prices and factor costs (Poole, 1983, pp. 50-52).

Bioenergy schemes could also—ironically—be crippled by competitive pressures on energy as a factor. An increase in demand for agricultural goods can be accommodated by more intensive cultivation, that is, by shifts from lower to higher value crops or by cultivation of additional cropland—changes at the "extensive margin." But at the extensive margin, energy inputs to agriculture become crucial since energy requirements of less productive land are greater relative to output. This type of expansion would thus tend to be self-limiting for export crops, including biofuels, because increases in farm commodity prices would reduce earnings per unit output due to elasticities in foreign demand. Furthermore, the increased total export earnings would be spent in disproportionate amounts for energy imports (Hertzmark, 1979, p. 1).[2]

The structure of farm management is also a subject of concern. In Brazil food crops are produced mainly by small landholders but the PROALCOOL program tends to favor large distillers with large plantations. There is some danger, then, of squeezing out the small farmers, forcing them into marginal land areas and thereby raising their production costs and utimately leading to higher urban food prices (Pelin, 1980, p. 25). In Brazil it was indeed found that the diversion of sugarcane from export to alcohol production affected domestic food production—especially on the most fertile land (Poole, 1983, p. 1).

2. In the context of the United State, the relevant question is whether fossil fuels like natural gas are being used as an integral part of bioenergy, either as process fuel or as a source of nutrients in fertilizers derived from natural gas.

The potential dangers of such a trend are obvious.[3] With a program goal of 15 million hectares by the year 2000, some see Brazil as developing a sugar-based monoculture (one-crop regime) (Pelin, 1980, p. 6). On the other hand, some analysts see this problem as overstated for Brazil and believe that—given some assumptions about a stable demand for corn, for example—the situation for land demand should remain stable in the near future. (Poole, 1983, p. 43).[4]

Suggestions have been made for formulating (or side stepping) the problem of calculating the costs of resource (factor) competition. One analyst suggests the following criterion:

> A set of policies designed to coproduce energy and food should proceed to the point at which the decrease in food export value is just matched by the replacement effect on energy supplies of the production of energy from agriculture (Hertzmark, 1979, p. 7).

Another suggests that the conflict can be ameliorated by (1) interplanting food and fuel crops (agro-forestry), (2) planting cash crops for energy purposes during fallow periods, (3) using only poor and underutilized land, and (4) employing waste biomass material (Hall et al., 1982, p. 148). The basic problem, however, is inescapable. The question is not whether bioenergy projects raise prices and costs, but by how much? In the most-studied case, Brazil, one analyst has indicated that rises in costs could be low—but only if a stringent set of conditions is fulfilled.[5] The planner must be aware of such complexities in order to make the best estimates possible of their consequences for project design.[6]

Pricing Policies

A major factor determining the degree of substitution is pricing policy, both for energy and nonenergy crops and for agricultural factors like labor. In an ideal market, if energy prices increase faster than food prices, resources will be diverted towards energy produc-

---

3. Problems with putting all the eggs in one monocultural basket are many: some biological, some economic. Besides efficiency problems, there are equity issues. For example, though production of energy crops raises aggregate farm income, it may lower income elsewhere in the society by raising farm-product prices (Rask and Adams, 1979, p. 11).

4. But note that Poole sees, in addition to possible underestimates of corn demand, problems in handling the *rate* of land expansion (Poole, 1983, p. 43).

5. The conditions for low opportunity costs are: demand-limited output; high supply elasticity; high price demand elasticity or favorable impact outlook; good possibilities for productivity improvements; and a de-linking of alcohol from the rest of the agricultural sector—e.g., separate capital funds available only for alcohol (Poole, 1983, p. 49).

6. The problem of food versus fuel or other products is often worsened by weaknesses in agriculture policy. In Brazil three export crops (coffee, cocoa, peanuts) experienced an increase in price while cultivated land declined because of change in the agricultural credit policy (Poole, 1983, p. 19).

tion. Ultimately, in a free market, prices would equilibrate. But farmers in developing countries rarely operate in free markets. In India, acreage decisions by farmers are constrained by government controls— and the organizational patterns of user industries. The domestic price for molasses is fixed, while it is exported at world prices (Bhatia, 1981, p. 7). Nor are all domestic prices rationalized: sugarcane is sold in India at a minimum price fixed by the federal government while state governments establish their own price support levels (Bhatia, 1981, p. 8).

In Brazil, also, sugar producers only indirectly respond to world prices because the government sets prices. Changing price policies could well have a profound impact. If sugarcane, for example, expands as a result of price support policy into regions that show varied crop outputs, it could have severe social and economic consequences— which would be reflected in changes in the prices of all factors of production. Other crops that could be used for alcohol production such as manioc might be hurt because they might not receive the same incentives (Pelin, 1980, p. 6). Still other agricultural products could be forced out of production.

Theory coincides with intuition in allowing that significant impacts are possible; but in practice the production decisions of farmers in reaction to price rises (elasticity of supply) tend to be difficult to predict. In India import controls on sugar in the 1930s resulted in an increase in sugarcane acreage. Traditional sugar refineries were displaced by more modern factories to supply the increase in demand, which led to a further increase in production and an expansion of irrigation (Bhatia, 1981, p. 12). Furthermore, one study found that in one Indian state changes in sugarcane acreage were found to be positively correlated with the price of gur (unrefined sugar) as it became more profitable than its competing crop, rice. In other states, however, there were no corresponding shifts in acreage (Bhatia, 1981, p. 13). It was found that the impact of higher sugarcane prices on changes in cultivation patterns and therefore on rural income distribution was a function of three factors: (1) acreage and yields of cane in various sized farms, (2) employment opportunities in sugarcane versus competing crops, and (3) employment levels in fuel alcohol facilities versus alternative sugarcane processing facilities (Bhatia, 1981, p. 28). The decision by farmers to shift toward increasing sugarcane production was affected by their use of the gur by-product, their ability to exploit seasonal price fluctuations, their shares in irrigation facilities, and their existing contracts with the mills and distilleries (Bhatia, 1981, p. 28).

Even free market prices will be affected by one bioenergy parameter: the development of underutilized land. The use of marginal or frontier land for both food and energy production will increase the strength and incidence of price fluctuations (volatility) in agricultural markets because yields will be even more difficult to predict than usual (Hertzmark, 1979, p. 15). This is especially important in countries with an established commodity futures market. But even in the absence of a futures market, a "new lands" policy could contribute to a general uncertainty in accurate project costing and poses a special problem for price-setting and quota-making policies.[7]

Environmental Impacts

Environmental problems are by definition problems that have to do with an area rather than a site. As we have seen in chapters 5 and 7, problems of land use and water and air pollution can be complex. However, the vast majority of them—like the disposal of stillage from an alcohol plant—are problems that affect the neighborhood of the bioenergy facility or a disposal site and do not fall directly in the province of national planning. Yet we have already noted that some impacts such as the effects of changes in biomass cultivation patterns on watersheds may be strongly tied in with national priority goals. Some aspects of air pollution impacts may also be of national—or international—concern.

The fact that some impacts are subjects for national planning does not mean that they can be easily brought under national control. In fact, the decentralized nature of much of biomass cultivation and technology makes control that much more difficult—a fact that reemphasizes the importance of project design for preserving the environmental quality of regional watersheds and airsheds.

Watersheds

Bioenergy cultivation can cause either the acceleration or retardation of erosion, depending on project design. In either case, loss of soil is a problem that has been increasingly recognized in the United States, for example, as an important national issue: it has been calculated that 3 billion tons per year of sediments flow into U.S. rivers, with the consequence that 230 million cubic meters (about 300 million tons) are dredged annually from U.S. harbors and rivers to keep them navigable, at a cost of about $1 per ton (Pimentel et al., 1982, p. 12). Soil loss and transport is also increasingly recognized as a crucial

7. Pricing protection for a domestic bioenergy industry will often be essential. In Brazil the private sector was at times reluctant to undertake substantial capital investment in distillery plant capacity, knowing that their production costs would still be higher than the cost of imported oil (Rask and Adams, 1979, p. 2).

problem in developing areas, where the loss of reservoirs due to siltation presents a direct threat to irrigation water for agriculture and the generation of hydropower. For example, the Peligre Reservoir in Haiti has experienced accelerated siltation, so that the present useful capacity of about 375 million cubic meters is to disappear entirely within the next 30 years or so (World Bank, 1982g). This represents, in concrete terms, a loss in electrical capacity of about 40 megawatts. Such problems are patently of national scope and concern. Furthermore, they are very difficult to control, because traditional farming customs such as the failure to retain residues in place may exacerbate them. Because such agricultural customs may have a rational basis in the short-term (OTA, 1980, vol. I, p. 119), enforcement of environmental rules would appear very difficult.

How does the bioenergy project planner fit into this picture? At a minimum, the bioenergy planner can recognize that particular bioenergy demonstration projects may well represent an "introductory sample" of future technology in cultivation and conversion. In that case it is important, regardless of existing "dirty" modes of operation in the traditional biomass sector, that new projects be designed to be consistent with perceived national goals. It may be supposed as that time goes on, national programs such as that in Korea (FAO, 1982a, p. 56) will address both water-related environmental goals—like the stabilization of hillsides and the prevention of erosion—and bioenergy priorities.

Airsheds

The impacts of most air effluents, as discussed in chapter 7, represent localized phenomena. On the other hand, the contribution of particulates to air pollution in typical rural villages has reached apparently significant levels of risk under unfavorable weather conditions, as evidenced by the palls of smoke commonly surrounding many rural villages at mealtimes (Smith and Aggarwal, 1983, p. 2358). It has been suggested that such pollution problems could be taken to influence the choice of bioenergy initiatives: the control of effluents in national airsheds could probably be more cost-effectively implemented by dealing directly with the problem of efficiency from traditional household cooking, rather than making large investments in expensive pollution control for modern biomass conversion units (Smith and Aggarwal, 1983, p. 2360).

Another "airshed" problem is apparently a local one: the role of vapor emissions or liquid contamination from methanol fuels in causing blindness and death (see chapter 7). Although the results appear local, the problem can be a national one, in the sense that if methanol absorption or ingestion is cumulative—a fact that, however, has not been established—the methanol project planner would not be justified in assuming that effluents based on one project will be a good guide to making a project compatible with national health standards.

Finally, there is a problem of national, or perhaps international concern, the question of climate modification. Predicted impacts of climate changes will be delayed and will probably take some time to manifest themselves; it is also a problem characterized by many scientific uncertainties; finally, it is a problem on which impacts would appear gradually and would affect different regions of the world in different ways (see for example, Ramsay, 1979, pp. 86-89).

Nevertheless, the question of climate modification represents a societal risk of possibly huge proportions. In one sense, the use of bioenergy can be taken as alleviating this problem because the most publicized effluent contributing to this problem, carbon dioxide, is reabsorbed by growing plants, roughly to the same extent as it is lost by burning plant tissues as fuel. There are, however, still two problems for bioenergy. One of them is that the change of biomass cultivation may not leave the carbon dioxide emissions in equilibrium. Clearing of natural forests and residue removal, together with replacement by plantations of short rotation tree or field crops would lead to a net loss in carbon stored in plant tissues. Increased water runoff would also contribute to carbon depletion, possibly leading to permanent decreases in the amount of carbon stored in soils (OTA, 1980, vol. I, p. 56). The second problem is that bioenergy in many forms, particularly conversion by direct combustion, leads to relatively large particulate emissions, which can also affect the climate. The problem is complicated by the fact that particulate emissions can either cool the atmosphere off—as has happened with eruptions of volcanos such as Krakatoa—or heat it up (Ramsay, 1979, p. 88).

This kind of question may appear very exotic to the average bioenergy planner, who will have much more to worry about closer at hand. Indeed, there is some question as to whether national policies on this problem can ever be coordinated—if scientific concerns are proven and the need for action actually arises. Climate modification, while possibly a real problem in the scientific sense, appears therefore to pose only a marginal concern for the bioenergy project planner at present.

Social Effects of Bioenergy Development

The impacts on social relationships of a bioenergy project are among the most difficult to assess and interact in a complex and subtle way with the social realities underlying national planning priorities. This is particularly true when the proposed energy project is to be introduced as an added element into the sometimes fragile homeostasis of a rural subsistence economy. Typical social issues from the project point of view could be how the project will be accepted in its initial stages and how the project will become integrated into the rural community. From the national point of view, such issues may represent typical characteristics of a nationwide syndrome: the clash of economic development imperatives with the parochial but complex lifestyles of traditional village economies.

There are three aspects of the social problem that can be considered in energy planning: equity, land tenure systems, and labor availability and population trends.

Equity

The most profound social impact of a bioenergy project could be in the form of equity, or changes in income distribution; this problem has been touched on in several places in this book (see for example chapters 8, 11, and 12). Although the society as a whole may perceive a net benefit, shifts in income distribution may occur that leave some poorer. This shift is often most apparent at the local level, but can also be seen at a regional or national level. For example, regions with better irrigation potential than others may benefit more (Bhatia, 1981, p. 29). Is there some generalization that can be made at the national level that could be useful for the bioenergy planner?

It seems very difficult to derive reliable guidelines. We have seen that small or "intermediate" technologies may have positive effects if leveling income differentials is the goal. At the same time, as with any type of economic development project, the equity effects of bioenergy programs are impossible to predict. The income distribution effects of a new technology are obscure in themselves—does more money go to managers on the average, or to workers, and what do we comprehend by "more"? Additionally, existing structures of price controls and subsidies, which often are designed as income supplement measures for the poor, will interact with new bioenergy programs; although these measures may not be the best in a theoretical sense, they may represent the only practical equity policies in many political and socioeconomic contexts (Dunkerley et al., 1981, p. 126). In practice, then, the bioenergy planner must be prepared for sharp differences of opinion and little hard data in this especially thorny area of development economics.

Land Tenure Systems

Commercialization of traditional biomass fuels could mean that the poor would become poorer and the rich, richer. In particular, land privatization could deprive poorer members of land for grazing or fuel collection (Hall et al., 1982, p. 131); this is a national-scale problem in many countries. As mentioned above for Pakistan, a cycle of partitioning, then appropriating, and finally privatizing community land has been discerned (Cernea, 1980, p. 16). Because domestic food in India is grown by small farmers and sugarcane is produced by large distilleries on their own plantation, alcohol bioenergy projects can produce land tenure changes and a reduction in the production of foodstuffs (Bhatia, 1981, p. 25). A similar situation has been noted for alcohol production from cane in the traditional prosperous agricultural states of central Brazil; the alcohol program led to an increase in the demand for land, effectively restricting land access for small landowners (Pelin, 1980, p. 5). As the demand for alcohol increased, land had to be zoned to inhibit food-producing areas from being converted to sugarcane (NAS, 1983, p. 86).

The differing reactions of large and small farmers to bioenergy development are relevant to the land tenure problem, but they cannot be easily predicted. In Pakistan, small farmers appeared more prepared to contribute to the costs of fuelwood plantings than did larger farmers: a firm contractual basis was essential in inducing small farmers to accept the risk of participating in a fuelwood project (Cernea, 1980, pp. 19-20). In other areas, it was the large landowner who seemed more interested. Larger farmers may have found it more beneficial to shift to planting more of an energy feedstock like sugarcane as a result of government support for bioenergy because they are able to take fuller advantage of government aid and infrastructure support.

As noted above, an analysis of land tenure systems must take into account both customary law and formal law. In practice, this may mean that new legislation or regulations will be needed. To make planning effective, it may become necessary to develop a general land use code at the national or regional level. This would require a codification of existing customary practices, including the development of sanctions to prevent practices defined as "misuse" in accord with national priorities. Bioenergy planners might have to be prepared for the impact of such measures.

Labor Availability and Population Trends

Labor availability and population trends may affect and be affected by bioenergy planning. As we have seen, in many societies women and children are the primary collectors of fuelwood. If charcoal is introduced to replace wood, men may supplant women and children as fuel producers. Also, a new class of workers may be created: those who work in the transportation of the charcoal to market intermediaries (Noronha, 1980, p. 27).

New inducements may be required to make labor available for bioenergy production. Existing migration patterns, for example, from rural to urban locations should tend to reduce the supply of labor available for biofuels production. On the other hand, a shift to plantation agriculture may displace workers, contributing to off-farm migration. In fact, often bioenergy programs will be explicitly designed to be labor-intensive and so help reduce rural unemployment or underemployment.

Planning Procedure: Interactions with National Economic Planning

At this point in the analysis, the relationship of a bioenergy option to questions of land, species, conversion technology, and project organization—together with modifying constraints like foreign exchange, infrastructure, etc.—will have been investigated. These factors, together with project-dependent factors like detailed costs and technical requirements, will determine its "plausibility"—i.e., general acceptability and prospects for success.

In most cases, this individual project as an economic phenomenon will occupy only a small place in the national scheme of things. Nevertheless, as a harbinger of what is to come, a new type of project may be viewed as quite important by national planners or interest groups. For example, the project planner will probably be aware of or will have brought to his attention such problems as "food versus fuel" or national land tenure policies. The main lines of most such problems—such as the true marginal long-run cost of land—will normally be comprehended in the analysis accompanying chapter 8. Some project design modifications, however, may be thought desirable after considering the wider scope of the social and economic problems involved in general national economic development planning.

Conclusion

There are certain areas in national planning that will interact very strongly with bioenergy projects. Because most biomass energy is involved with plants, it comes as no surprise that the interaction with the agricultural and forestry sectors is of the utmost importance. Because of its greater economic value per square meter, the agricultural problem is foremost. The controversies surrounding using field crops for bioenergy purposes—usually referred to as "food versus fuel"—are complex. The question is examined here in the context that there is necessarily an opportunity cost of some kind in using bioenergy. In some cases, this cost may be large. However, there is good reason to believe that the growth of bioenergy will not be fatally impeded by land constraints; it may, however, be constrained by the lack of scarce factors, management and capital, as well as by market indivisibilities of a "social" kind.

Environmental concerns, while a part of project planning, also have strong connections with national planning. The role of bioenergy in modifying the parameters in the climate modification problem is perhaps an extreme example. But the importance of bioenergy plantations and other cultivation and conversion facilities in planning for the protection of watersheds for agriculture and hydropower is evident. Finally, the "social" problems involved in equity (income distribution), land tenure, and labor supply touch directly on national priorities. Although evidence on the role of bioenergy in these concerns is as tenuous as in any other economic sector, it seems reasonable to guess that intelligent bioenergy planning can provide an opportunity for social amelioration.[8]

8. Much of this chapter is based on drafts by Richard Kessler.

14

Bioenergy on a National Scale

Introduction: Programs That Make a Difference

The nature of national policies toward bioenergy can make or break the individual project. Sometimes there will be no official policy on bioenergy. But increasingly governments will have some kind of official attitude toward renewable energy development, biomass resources, and toward biomass energy in particular. This attitude will be important enough for the individual project but will be crucial for the future of bioenergy in the country.

Therefore, the development of country-wide programs comprising a number of projects—the logical next step in making the bioenergy a practical energy option—will hinge in part on existing national policy. Whatever that policy, however, successful development of national programs starting from a base of local projects is a complex and little understood process. This process requires continuing investigation of past experience and imaginative planning for future contingencies. Here, the planner is truly dealing with decisions that "make a difference."

Agency Structures and Interests and National Policies Toward Bioenergy

Agencies in some countries may have special policies and programs in the bioenergy area. In such cases, the individual bioenergy project may have its directions predetermined by existing national policies. At the very least the planner may have to devote a good deal of effort into fitting in with existing national energy goals.

The Institutional Multiplicity Problem

It is apparent from the name alone—the words "biomass energy" suggest biology (forestry, agriculture), as well as energy—that the bioenergy planner may have to deal with a number of different government agencies in planning and implementing the project. Projects such as eucalyptus plantations for fuelwood or sugarcane for fuel alcohol production will touch on the interests of a ministry of energy, a forestry department, and a ministry of agriculture—and perhaps several others. As we have already seen (particularly in chapter 13), there may be real conflicts between bioenergy and other development goals—such as the use of land for production of biofuels instead of food and fiber. On the other hand, possibilities for fruitful cooperation between energy agencies and other departments may arise such as the use of forestry plantations for both fuelwood and for preserving watersheds to protect agricultural production. It is, therefore, apparent that agencies dealing with energy, forestry, and agriculture may have to cooperate to resolve conflicts involving bioenergy, and that often conflict can be avoided by agreeing on bioenergy schemes that achieve multiple energy and nonenergy development objectives.

How can the bioenergy planner go about securing such cooperation? In many cases this may be a difficult task. For one thing the agencies involved may have somewhat narrow, parochial conceptions of their duties. This narrowness may mean that bioenergy projects will either not be comprehended or will not be seen as falling within the sphere of interest of the particular agency. If that happens for two or more agencies, one can see that the planning for a given bioenergy project—for example, the use of water hyacinth in a methane fermentation digester—could fall into an institutional "gap."

Such gaps often arise through historical accident because of the haphazard ways that agencies may have been formed. An energy department may have little understanding of renewable energy in general and of bioenergy in particular.[1] Forestry departments may be primarily concerned with conserving existing forestry resources or with the promotion of industrial forest projects—sawtimber and pulp—not energy.[2] Agriculture departments may be understandably concerned with spreading or consolidating the results of the green revolution, achieving goals such as self-sufficiency in grain production, or establishing new export crops. The use of field crops for energy, the use of trees together with crops in agro-forestry schemes, or the use of crop wastes for gasification may be too far removed from usual departmental concerns to deserve attention.

1. Often energy departments may have been created from or appended to petroleum or mining departments or to national electrical utilities, which will be used to thinking in terms of fossil fuels or hydropower instead of biomass.

2. While there are many exceptions, and more every year, many forestry departments have seen themselves as protecting the forest from the unauthorized use of trees as fuelwood (poaching) rather than as promoting sustained fuelwood production and use. In many cases, furthermore, forestry utilization for timber products has not been carried out on a renewable basis. Both short-term and long-term economics usually point to renewability ("sustained yield") as a requirement for forestry as a source of biofuels.

Such a deficient institutional structure is by no means universal but on the other hand is not uncommon in the Third World. Its consequences are that it may be very difficult to resolve bioenergy conflicts through existing institutions. Worse yet, the cooperation needed in implementing projects that could help forward the goals of two or more government agencies may not be forthcoming. Sometimes governments and even private donors have tried to override such problems by creating new institutions or by taking special steps to secure the cooperation of older institutions. As we shall see below, such cooperative programs, although they can be effective, may have their problems along with their benefits.

Possibilities for Cooperation and Conflict Resolution

The existence of institutional gaps means that it is especially important for the bioenergy planner to be able to identify possible compatible goals that could be achieved through a project, or at least to recognize and try to adapt to the conflicts arising when bioenergy goals trespass on other stated development priorities.

Complementary Goals and Cooperative Efforts. In some institutional environments, it is especially important for the bioenergy planner not to restrict himself to the considerations of a cost-benefit analysis based on narrow project goals. In many practical situations, the most important factor in securing the success of a project may be its compatibility with the competing goals of several institutions or constituencies.

Biogas, for example, can often be combined with other goals, sanitation, for one. In China, the official propaganda for methane fermenters has stressed the fact that fermentation carried out over a long period kills dangerous bacteria and viruses and thus helps village sanitation (Chen and Chen, 1979, p. 112). This same compatible goal has been sought in a free-market setting in the Philippines, where biogas production has also been introduced as part of the sanitation scheme for large pig farms (Terrado, 1981, p. 146). An even more familiar side benefit of biogas is that it preserves fertilizer instead of consuming it. Animal waste (and vegetable waste) contain, along with carbohydrates, important nutrient elements such as nitrogen and phosphorus that are lost to the atmosphere when wastes are burned but are retained in the residue (slurry) from the biogas fermentation.

The examples just mentioned are ideal in that two different jobs are accomplished by using few or no extra resources. More often than not, however, valuable land, labor, management, and capital resources must be added to the bioenergy scheme if it is combined with other goals. A good example is national reafforestation schemes, in which goals of wood production both for fiber and pulp and for fuel have been combined with the preservation of watersheds. Bioenergy planners can usefully study the example of Korea, where a forestry program was subsumed as part of a more general "New Community Movement" and where a great deal of successful reafforestation was accomplished (FAO, 1982a, passim).

In the Korean case, conflicts inevitably occurred and often left their mark. Special rules had to be set up to separate the spheres of agriculture and forestry: the Korean rule of thumb required that all slopes over 35 degrees be forested rather than planted in field crops. At times some goals were met while others were not. For example, by 1978 only half of the acreage for erosion control in Korea had been treated as planned (FAO, 1982a, p. 70). But for the program overall, new ideas were often successfully realized through compromises and adjustments in traditional institutions.

When large commercial markets are at stake, the interaction of bioenergy goals with other economic targets may become truly complex. The first development of fuel alcohol in Brazil, for instance, was based on efforts to dispose of sugarcane crops during years of depressed international markets (Carioca, 1981, pp. 2-3). But a doubling-up on facilities, having one set for making sugar and another for distilling alcohol, is not cheap.[3] A program of pure fuel alcohol has also boosted sales in the Brazilian automobile industry, because modified vehicles have to be produced for running on pure alcohol fuel. Obviously, using bioenergy programs to stimulate artificially other industries could play a role in a national development plan—but the cost can be high.

Inevitable Conflicts. The bioenergy planner will inevitably run into conflicts that are too difficult to reconcile with the goals of other sectors. One very common example has to do with irrigation water policy in arid or semi-arid regions. The use of irrigation can greatly increase the yield of such species as eucalyptus in semi-arid zones. Nevertheless, meeting agricultural goals for crops like rice and wheat may be a much higher national priority. This obvious kind of trade-off is sometimes neglected by official policy and only makes itself known in practice to the planner who sees his eucalyptus becoming stunted while water is diverted away to nearby rice fields.[4]

A related difficulty: short-term goals are often able to steal priority from long-term targets. For example, we have noted that a fuel alcohol program in Costa Rica could help the national balance of payments in

3. The expense comes from the extra opportunity cost: while sugar is being produced in the sugar refinery, distilleries lie idle, and vice versa.

4. A similar situation appears to have occurred in places like the Sudan (Sudan, 1982), where there are both great needs for fuelwood and very high priorities set on agricultural export crops.

the long-term. But it could cause a fatal drain on foreign exchange in the short-term as distillery technology, farming equipment, and other foreign capital goods were imported to start up the industry (Celis et al., 1982, pp. 19-24). The short-term foreign exchange crunch has in fact been the dominant factor there up to the present.

Informal one-sided conflict resolutions may involve such high priorities that the consequences for a given bioenergy project will have been obvious from the start. On the other hand, the conflicts can be fairly subtle in nature and worth watching out for. Such informal resolutions of conflicts could conceivably be labeled crisis management, or policy by default. For the practical planner, however, it is usually more comforting to think of them as ways of settling intractable socio-economic problems through the only practical route available, the political process.[5]

National Coordination on Biomass Energy

The approach of any one nation to bioenergy may be fragmented or nonexistent. Conversely, it may be rather well structured into different modes of organization.

Nonexistent or Formalistic Coordination. Perhaps the most common circumstance in most developing countries is that national bioenergy policy is undefined, and a minimum of coordination will exist between the various agencies that by rights should be concerned with bioenergy planning. This policy vacuum can be expected to strengthen the impact of existing institutional gaps on new initiatives.

Such an environment could be fatal to bioenergy planning—whenever cooperation is indeed essential to success. On the other hand, it sometimes opens the field to initiatives by individual agencies or to cooperation at lower levels of agency hierarchies. For example, the Khadi and Village Industries Commission—set up originally to promote the weaving of homespun cloth—has played an important role in the development of the biogas option in India. This development of an an important bioenergy technology, involving a current total of over 100 thousand users, illustrates the possibilities for a bioenergy program carried out on a (primarily) single-agency basis (KVIC, 1983, passim).[6]

The bioenergy planner can also usefully study examples of the growth of agency coordination from the bottom up. In the State of

5. The underlying socioeconomic problems might well take the form of a high discount rate, making the present much more important in the future, or they might be seen in the form of larger financial interests tied up in field crops compared to forest crops. Or they could involve the basic political power equation of using resettlement schemes, for example, as weapons in a war against internal political dissension.

6. The biogas program is not entirely a one-agency show. Various state government agencies have also played important roles and the KVIC work has been duly scheduled as part of the national Five Year Plans.

Gujarat in India, informal cooperation between the Forestry Department, the Agriculture Department, and the local village (*panchayat*) organizations has been an important part of a successful tree planting program. Technical exchange relationships such as special training courses by the Forestry Department for Agriculture and *panchayat* personnel undoubtedly helped cement local-level cooperation (Java, 1984a).

Donor Guidance. The typical bioenergy planner may be dependent for funds and perhaps technical assistance on one particular donor or on several. Some donors play a passive role in country situations. In other cases, however, individual large donors or groups of donors may come to exert an influence that makes them almost a shadow planning agency for bioenergy or other development projects. This situation is apt to have even fewer understandable ground rules than more formal organizational hierarchies. The bioenergy planner must therefore be conscious of the typical faults of donor operations—such as lack of continuity and cultural incompatibility with the local society —as seen, for example, in cases in Burkina Faso (Upper Volta) where forestry projects failed to catch on (Hoskins, 1979, p. 7). On the other hand, combinations of donors acting in concert— such as has been reported in forestry programs in Nepal—offer a creative example of using donor staff capabilities to provide organizational leadership for bioenergy programs (FPL, 1980, pp. 67-68).

Blitz Programs. Some countries have mounted serious nationwide campaigns in support of bioenergy programs, usually forestry efforts. Large national programs have been organized, agencies reshuffled, local institutions pre-empted, and mass media campaigns undertaken.

Bioenergy planners caught up in one of these "blitz" campaigns may well find many obstacles in their path smoothed in the course of carrying out their own small part in a national torrent of effort. Certainly, in the case of Korea, new projects financed by World Bank loans were able to fit well into the already established program of the Village Forestry Associations, in concert with the more general "New Community Movement" dedicated to economic, environmental, and even spiritual regeneration.[7]

7. The Korean program was characterized by a number of new institutions called Village Forestry Associations (established in the Forest Law of 1961), county-level Forestry Association Unions, and the National Federation of Forestry Association Unions. They in turn were associated with local agricultural cooperatives and with the entire "New Community Movement" (Saemaul Udong) and its institutions, with various tie-ins between government agencies at different levels. (For "New Community" goals, see FAO, 1982a, p. 24.)

Media coverage was also a key factor, as were probably some key institutional changes such as transferring the Office of Forestry from the Ministry of Agriculture to the Ministry of Home Affairs, the agency controlling the police force (FAO, 1982a, pp. 12ff, 34-36).

On the other hand, a blitz program organized in Tanzania succeeded in its goal of establishing a large number of village woodlots, but the woodlots in general were too small in size to make much difference in meeting national fuelwood norms (Skutsch, 1983, pp. 50-51). It is not clear whether individual donor-financed projects carried out within this national program tended to be any more successful than usual, though programs run directly by local schools or churches were (as often, elsewhere) more successful than the average and seemed to flourish in the blitz atmosphere.

Economic Juggernaut. In centrally planned economies, the "blitz" or partial blitz approach is often combined with special amendments to the laws of economics in order to expedite development efforts like bioenergy projects. This is undoubtedly true in both the biogas and reafforestation programs in China, although the record is complicated by periodic changes in political outlook and by uncertainties in reporting and accounting (Taylor, 1981, pp. 141-142, pp. 211-214). Nevertheless, it seems that the forestry program's success owed a good deal to the complex organizational framework that reached from national levels down to the humblest village—and could call on the local police power (Noronha, 1980, p. 9).

The re-working of national economics in order to ensure a successful bioenergy program is best illustrated by the PROALCOOL scheme in Brazil to replace oil products by alcohol from sugarcane and other feedstocks. The Brazilian program certainly involves the same elements of media coverage and institutional changes seen in Korea and Tanzania: for example, the reorienting of the marketplace through new regulations by both new and existing government agencies has indeed been dramatic.[8] The result of such efforts may ensure a reliable supply of biofuel—but the question of "economic feasibility" tends to become buried in a vast network of price fixing, quotas, subsidies, and export controls.

The Development of Country-wide Programs

Most planning efforts in bioenergy will go into the design and execution of a single project or a small group of projects. This book deals with such projects, that is, with bioenergy cultivation and conversion facilities of a restricted size and usually in a localized area. But there is no overriding reason to restrict the treatment to local projects. We have seen biomass programs in Korea, for example, which take a completely national and comprehensive point of view, with many small

8. See chapter 10 for details.

(and large) projects participating. However, the project approach is usually the sensible one: funds are restricted, technologies are unfamiliar, and management is not in long supply. Indeed, problems in some of the bioenergy projects in Panama, for example, may have arisen just because of the complexity of the projects, their inclusion of the multiple goals of watershed protection, soil conservation, introduction of improved stoves, and tree planting on a quite massive scale (Thrupp, 1981, p. 26). Often, indeed, it is only by the learning experiences that come from some pilot or demonstration projects that one can sensibly plan large programs.

One project, even of a fairly substantial size, however, can usually make very little impact on the total energy situation of a particular developing nation. As has been emphasized repeatedly above, the responsibility for anticipating the directions of national development programs cannot be expected to fall on individual project formulators and evaluators. On the other hand, as we saw earlier in this chapter, the project planner cannot afford to ignore national planning. In however small a way, one particular bioenergy project will both be affected by and in turn affect the total national picture. Furthermore, the design and execution of individual projects should take into account certain basic facts and problems having to do with the possible role of bioenergy in the economic development of a nation.

For example, realistic goals for bioenergy projects in a nation where the total bioenergy potential seems very small will usually be quite different from that of a nation where the potential for biofuel seems very high. Furthermore, as we have already seen in chapters 2 and 8, alternatives to bioenergy—oil and other fossil fuels, especially, but also such renewables as hydropower—will play an important role in the economic desirability and viability of a particular project. Since we generally want to look at one project as a precursor to other projects, it is important to understand how the project can or cannot be replicated—that is, imitated or reproduced—in other places. The planner of projects will have to adapt to the institutional infrastructure of a nation just as a fish must swim in the sea; but the planner will also have liaison with, and exert influences on the local institutions of a country. Finally, there is the question of monitoring and evaluation—how can one keep an eye on the projects and make certain that they are doing what they were designed to do?

Realistic National Bioenergy Goals

There are two different ways of approaching potential goals for a national bioenergy program. One can concentrate on the fact that bioenergy in the form of traditional fuels is already a very large sector in many countries. Even in Latin America, the continent least dependent on these fuels, about half of the population depends on fuelwood and charcoal for basic needs (see table 2-B-I). Or one can stress the other aspect of the problem that is receiving most emphasis in this book, that is, the use of modern biofuels to replace present commercial fuels or substitute for traditional fuels in some end uses.

The practical problems, however, are not much different whether or not a country already has a large traditional biomass energy sector. In either case, the planner is faced with complex problems in estimating the potential of biomass resources—the forests, wastes, and energy field crops that provide the basis for traditional and modern feedstocks. The difficulties of making such a resource assessment have been pointed out in chapter 3 on land use and in chapter 4 on cropping. In chapter 13, as well, we have seen how the development of bioenergy feedstocks on a significant scale may depend on finding a fit to the multiple-use objectives of policy at a national level.

To the degree they affect his project directly, the planner cannot avoid facing the problems discussed in those chapters—such as the unreliability of land use and forestry data and the ambiguous status of crop wastes (which may or may not be "waste"). One basic problem is to assess the overall "favorability" of the outlook—roughly, the long-term supply prospects—for bioenergy in a particular country. This outlook is, in the example we have cited before, much more favorable for Brazil than for Egypt. Both have unused land that is submarginal under the mix of factors of production presently employed. The important distinction between the two country cases is that the use of new land in Brazil depends on the development of new infrastructure, particularly roads, and this new infrastructure may come to make economic sense in the future. The key factor preventing the use of more land in Egypt, however, is water, and the opportunity cost of water will probably continue to be high indefinitely.

Other constraints may come in on the demand side. A country such as Costa Rica has promoted the use of diesel fuel rather than gasoline. Because diesel is less easily replaced by alcohol, such a policy lessens the desirability of alcohol fuels programs.

Incidentally, information gaps can make a difference between countries. Crop wastes in Rwanda, for example, may—or may not—be more available for energy use than in, say, India. The advantage in India with its relatively good statistical data base will be that the planner will have more adequate information available documenting the widespread utilization of wastes for soil amendments, fertilizers and for construction materials, as well as for traditional fuels.

Ideally, the kind of summary analysis suggested here should be sufficient to show how a particular project fits in rationally with overall goals appropriate for a bioenergy program—even if no bioenergy program is as yet in place. A good example of investigations that can be helpful in this regard is a recent study done for direct combustion, gasification, and other wood-based technologies in Costa Rica (See Meta, 1982, passim). But even in default of such a study, the planner should be able to outline plausible scenarios for a national bioenergy program as a point of reference for project planning.[9]

Alternative Energy Sources at the Macroeconomic Level

The present demand for fuels other than biomass energy (treated in chapter 2) and the costs of alternatives such as oil and hydropower (treated in chapter 8) will affect the viability of the project. But the competitive situation vis a vis alternative fuels will also be a primary determinant of the outlook for larger bioenergy programs that involve numbers of individual projects.

There are two related problems when trying to look at the bioenergy picture in relation to the competition from alternatives such as oil, coal, or even wind energy. One problem is that the energy markets in which biomass fuels will compete are already very complex. Factors like regulations, price controls, subsidies, and environmental standards often make a comparison between fuels difficult (OTA, 1980, vol. I, p. 142). In an ideal world, the planner would be able to rationalize all these different constraints and to establish a sensible system of alternative fuels—including but not limited to biomass—that could be the basis for a national energy policy.[10] In practice, the average bioenergy planner will have to take such complex policy instruments as the current Indonesian subsidy on kerosene as a given factor in cost calculations; the only job is to understand the extent of the subsidies and how they may affect his bioenergy decisions. This task is not always easy. It might have been difficult to predict in 1973, for example, that subsidized kerosene would come to be used in some countries as a replacement for diesel fuel in farm machinery.

9. When modern biofuels become an important source of a nation's energy supply, the country can become a hostage to fortune as far as dependence on weather or on pests that may especially threaten plantations in monocultural (one-crop) cultivation. It has been suggested that when biomass comes to supply a certain percentage of a nation's energy supply, it might be relevant to stock up on biofuels, making a kind of "strategic reserve" that could buffer some of these dangers (OTA, 1980, vol. I, p. 143).

10. One reference (OTA, 1980, vol. I, p. 142) lists several steps that "should" be used to form biomass policy. These suggestions are laudable, but they could imply actions that would be difficult to implement. They involve reviewing policies that currently favor, directly or indirectly, fossil fuels and nuclear energy, etc.; reviewing policies that subsidize or discourage biomass fuels or byproducts; and implementing policies for information dissemination and commercialization of bioenergy.

The other cost comparison problem arises from possible complexities in the market for bioenergy itself. That is, biomass use could be promoted through subsidies or by administered prices that could succeed in obscuring what the costs of alternatives actually are. The planner might instinctively welcome subsidies for the development of bioenergy. The problem with subsidies, however, for those with vested interests in bioenergy, is that they may be here today and gone tomorrow. And the future loss of a bioenergy subsidy could mean not only change in the marginal economics, but also the entrapment of a good deal of valuable capital in uneconomic bioenergy plants and equipment that would be no longer competitive.

Usually the first problem, that of subsidized alternatives like kerosene, will have to be dealt with in the project planning itself, as indicated in chapter 8. Regarding the second problem, policy for bioenergy may not be in place at the time of the implementation of pilot projects, so projections of bioenergy subsidies may be impossible to make. And hypothetical subsidies that cannot, in any event, be guaranteed, should be excluded from the analysis. Therefore the net effect of future changes in competition from alternative fuels on programs (groups of projects) versus projects should in practice be small.

Impediments to Replication or Extension of Projects

Developing an entire program of biomass energy can be a major exercise in national planning. Such programs usually require action at the highest political levels and involve major decisions about the use of public funds as well as the participation of private industry and sources of capital.[11] At this level, the choice of development strategy for replicating a particular project therefore can be quite complex. Compromises may have to be made between project designs that seem to "standardize" well from a technical point of view and those that happen to be socially or politically acceptable.[12]

The individual project planner does not often have time to think about nationwide problems that affect replication, nor is he briefed to deal with them, but he has to be able to deal with and understand the context into which his project fits. A good deal, moreover, can be accomplished at a lower level. Certain basic support functions can

11. It has been pointed out by Hirschman (1967, pp. 123-125) that it is very difficult to achieve the right mix between private and public outlays, especially in the areas where public funds tend to be tax-starved. The public and private sector each have a tendency to incur the least possible cost with the result that insufficient services are provided to the public.

12. Certain types of projects—independent of their intrinsic nature—may be more desirable than others: for instance, projects connected with particular sites allegedly have an intrinsic political advantage over "non-site-bound" projects (Hirschman, 1967, p. 78).

determine whether replication will work or not. For example, evidence from one study of past AID projects and programs suggests some defects in program support that might be avoided (Ward et al., 1983, [TAB A1], pp. 4-5). In particular, the study found that the marketing and dissemination of systems and equipment in many AID projects in Africa were ineffective; deficiencies in this key area could prove a severe handicap in project replication. The most salient problems had to do with a lack of systematic testing of markets, a failure to use advertisements, and a lack of project personnel experienced in business enterprises.[13]

One of the prime factors involves what could be formally termed "management," but which in practice has to do with the problem of to replacing the innovative personalities that often make a demonstration project successful. To go from one initiator (or more) to reach numbers of communities usually takes organization. The large farmers—in the case of fuelwood—often have to be reached. But experts have noted that "organization can kill," and the task for the national bioenergy planner in establishing viable but flexible organizations is not easy (Noronha, 1980, p. 24).

Often the best way to organize for good management is to use established organizations. This has been pointed out (in chapter 10) as an advantage of bioenergy options that use byproducts of existing industries. This approach has also been useful in the Philippines dendrothermal program, in which the skills of the staff of existing electric cooperatives can be tapped not only for the utility operations but also for supervising finances for, and giving guidance to, the Tree Farmer Associations (Denton, 1983, p. 24).

There are other general problem areas that are usually relevant to replication such as the question of repair and maintenance.[14] There are also problems that have to do in particular with the design of renewable energy systems, or more properly speaking with small scale, "intermediate technology" renewable energy systems (see chapter 8). For such systems, it has been alleged that the planner must be unusually sensitive to the reactions of people because the social aspects of the technology play a prime role in its planning (Thrupp, 1981, p. 6). When replication—building up from individual projects into larger groups of projects or programs—is being carried out for such intermediate technologies, the importance of the participation of social

13. The AID study expressed the belief that many projects had overlooked obvious possibilities for supporting private sector marketing of already commercially available equipment.

14. Some recommend (Ward et al., 1983, p. 6) that equipment of high durability and low maintenance requirements be chosen, others that "continuous maintenance" is a surer course. In either case, however, emphasis in any large program must logically be placed on technical support training.

scientists, for example, has been stressed (Soedjatmoko, 1979, p. 64). This aspect is particularly important when the technology, as often, is to be applied in subsistence economy environments.[15]

Some replication problems have to do with the nature of bioenergy, as distinct from other forms of energy, renewable or fossil. Bioenergy systems, as we have seen, can be large or small, commercial, community, or government-run, so the replication problems will vary depending on the exact systems chosen. As we have seen above (in chapter 8), special problems may arise as a function of scale. For example, it has been noted that it may be difficult for small farmers to replicate a gasifier project independently because they have no way to lease land, they cannot afford the amount of capital needed to set up the initial wood production facilities, nor can they afford the conversion costs for adapting the diesel engines for gasifier operations (AID, 1982, Annex II, p. 16).

Other replication problems arise from the multiple-product nature of biomass. In fact, one analysis suggests that the existence of non-energy markets makes it difficult to put into effect policies that apply to bioenergy uses alone (OTA, 1980, p. 142). This observation fits in with the discussion earlier in this chapter on the intimate connection between energy, forestry, and agricultural policies involved in bioenergy.

Many bioenergy technologies do not produce energy in forms that fit readily into existing energy distribution or consumption systems (OTA, 1981, vol. I, p. 142): this problem can often be fatal for a large program. This stricture applies to such fuels as pyrolytic oils, producer gas, and to some extent to biogas. This means that new, site-specific means of processing, distributing, and consuming the resulting fuels—such as building pipelines for biogas or adapting diesel engines to gasifier feed—often must be developed to make the bioenergy option commercially attractive on a large scale.

Finally, in addition to the obvious problems of infrastructure and supply, the planner must be on guard against apparently minor constraints that may have major impacts. For example, biogas systems not only require a supply of animal (or crop) wastes, but they also require water, a resource readily available in some areas and not in others. In the same category fall cultural factors such as attitudes towards han-

15. For example, one study has recommended that less emphasis in both budget and administrative allocations be given to large-scale isolated systems (for example, tree farms) and more to decentralized multiproduct systems (OTA, 1980, vol. I, p. 150).

It should not be thought, however, that bioenergy's programs should necessarily deal primarily with low income rural populations. although such populations may be a reasonable target for bioenergy projects of the intermediate technology kind, it has been noted that in practice most renewable energy systems have been rather capital intensive; the first to adopt most systems have been relatively affluent urban and rural dwellers, medium-to-large scale entrepreneurs, and social service organizations (Ward et al., 1983, p. 7).

dling wastes; again, biogas schemes that may be satisfactory in some social milieus may not be in others.

The point to be stressed is that these problems in replicating and extending individual projects can often be anticipated in the project design stage. The pilot or demonstration plants can then be modified so as to increase their chances of inspiring a successful replication program. One example arises in options for improved stoves. Stationary sand and clay stoves must rely on the use of extension programs for dissemination (Ward et al., 1983, p. 26). Portable metal stoves, on the other hand, in addition to other virtues such as uniformity and high efficiency, can be easily transported and introduced into existing markets.

Institutions and Liaison

There are several valid approachs to the problem of liaison between bioenergy projects and programs, government, and donor and local institutions. The bureaucracies that will deal with bioenergy problems may vary a great deal, as we have already suggested. But whatever official institutions exist, the bioenergy planner will generally have to maintain a close liaison with them—particularly if his projects should become involved in a nationwide program of bioenergy development.

The success of liaison depends not only on the structure of the hierarchy and on national political decisions, but also on attitudes and informal understandings. Often, despite official attitudes, enthusiasm for new types of energy systems like bioenergy will be lacking; this may hold true even if there happens to be considerable intellectual influence or political clout behind new energy options (Thrupp, 1981, pp. 7-8). Liaison problems will arise not only with governments, but also with international donor organizations, and interaction with such entities can be fully as demanding as with any government department (Thrupp, 1981, p. 20).

The liaison with the authorities at some levels may be complicated by the phenomenon already referred to: the built-in reward and punishment system characteristic of most bureaucracies. A typical "district officer" will have his pay, "perks," budgets, and upward mobility opportunities all flowing from his superiors; all incentives are upward and pertain to the goals of his agency alone (Wunsch, 1980, p. 22). It is difficult to attain a multisector approach or a sensitive response to community needs in such an atmosphere.

It is apparent, therefore, that liaison with institutions as part of the usual type of of top-down planning process may present difficult problems. Critics of fuelwood initiatives have noted, indeed, that top-down planning just "doesn't work" (Hoskins, 1979, p. 3). Be that as it may, if enough local interest can be generated in a community, one may be able to promote the alternative concept of bottom-up planning as part of the bioenergy liaison process. For example, a village seeking authorization to develop its own fuelwood project might have difficulty in many countries just because of the number of hierarchical levels involved. Decentralization of governmental planning agencies could alleviate such problems.

Advocates of participatory democracy have publicized the special value to the communities of encouraging the creation or fostering of new or existing decentralized institutions, especially in matters relating to the meeting of basic needs—including basic energy needs (Soedjatmoko, 1979, p. 58). This method has been tried in Korea and Nepal, for example. In one arrangement in Nepal, the panchayat (village council) decides whether or not it wants a woodlot. The District Forest Officer then examines the proposed lot for technical feasibility, sending the scheme on to the District Panchayat, which sets priorities for schemes from various communities (World Bank, 1980d, p. 41).

Thus, although it may be desirable to set up new local institutions, existing local entities may present an opportunity for greater efficiency in mobilizing human resources. In Malawi, failures to link the formal organization of fuelwood schemes were noted, while in Rwanda, the official schemes failed to connect with the traditional farmer associations (World Bank, 1980d, p. 42). It is difficult to generalize, but it appears evident that even obstructionism on the part of established institutions would best be dealt with directly rather than by outright neglect of established interests.

To be sure, the increased utilization of village institutions through decentralization may in some cases cause a drop in efficiency, owing to inexperience and lack of administrative skills. Believers in these methods, however, have stressed the values in improving self-confidence and self-discipline through institution building (Soedjatmoko, 1979, p. 65). Examples of this kind of change in institutions include an extension of the judiciary system to villages to help them weather increased stresses and strains on the legal system—like land tenure disputes—brought on by the processes of economic development (Soedjatmoko, 1979, p. 66).

The key to resolving social conflicts in developing bioenergy production is often found in liaison methods that bridge the gap between participants—consumers and producers—and management—government planners—including extension workers. At the national level only public costs and benefits are perceived while at the production level only private costs and benefits are seen. There is also a social as well as economic "distance" between government and village that may require special monitoring. At the village level the dominant motivator of innovation may be fear (Campbell, 1980, p. 7). In such a case, a government might need to establish its credibility to obtain community cooperation and reassure its people. The ambiguity inherent in other community values such as land ownership contributes to a system where user groups can misperceive each other's role and management responsibility may be lost (Campbell, 1980, p. 4). The abuse of customary timber rights and of concessionary issues can limit the role of a forest department in promoting fuelwood, putting the government in conflict with the population (Cernea, 1980, p. 3). The need to resolve these ambiguities can become a major concern of national energy planning.

There are then two sorts of roles that the planner may have to fill. On the one hand, he may have to perform a complex process of liaison with a well established bureaucracy, following pre-existing rules and negotiation patterns. At the other end of the spectrum, the planner may have a chance to participate in the building of new institutions, especially at the community level, and the establishment of new patterns of project liaison. In any case, it should be noted that the interaction of the planners both with communities and with government institutions is a two-way street, involving communication of reactions—not just actions.

Monitoring and Evaluation

Monitoring and evaluation—highly desirable activities to include in individual bioenergy projects—become indispensable if that project is to become part of a larger program. Monitoring is concerned with the examination of results while the program is ongoing, while evaluation is concerned with assessing the results afterward.[16]

A certain part of project monitoring has to be carried out for ordinary accounting purposes, ensuring that project management can keep track of recurrent costs, maintenance expenditures, and system outputs. A recent review of certain AID African projects identified some common monitoring problems. Performance monitoring on these projects was thought generally inadequate, and a particularly troublesome point was that monitoring methodologies were not compatible, so that it was difficult to compare results (Ward et al., 1983, p. 63).

16. One source defines evaluation in the following terms (Turner, 1979, p. 71): evaluation for projects is appropriately performed after external support of the activity has ceased; it is done by a detached observer; it examines the relevance and need for the project, its design and underlying assumptions; it assesses induced change and progress toward targets; it identifies unplanned change; it identifies causal factors and assesses their effects; and it feeds the findings into a redesigned and improved plan of execution.

The problem for groups of projects was worsened by the failure to pool data on field tests (Ward et al., 1983, p. 7).

Market and resource surveys to identify opportunities and consumers are a very desirable monitoring option (Weatherly, 1983, p. 1). Market surveys provide options to national planners by giving information about demand; resource assessments give information about the size of the sustainable resource base. Using these institutional tools, the project management can coordinate inputs of technical assistance, social and institutional organization, and financial arrangements to ensure that the production and consumption system functions smoothly (Weatherly, 1983, p. 2). The project management system can thus provide the link between local markets and national development objectives. For instance, the use of crop residues for bioenergy production is affected by the seasonality of supply and of competing demand patterns. A bioenergy program that does adjust to such constraints could affect many other economic sectors.

Advice has been offered to planners to establish progress indicators for monitoring by setting up criteria to provide an objective assessment of project progress (Turner, 1979, p. 75). Appropriate criteria cover a wide range of values. For example, many bioenergy projects have significant equity consequences and also impacts on other social dimensions. It is therefore important to be sure that observations on such topics as social acceptability are included in monitoring procedures (Ward et al., 1983, p. 64).

"Accelerated use" tests can be an important part of monitoring. None of the countries visited in a recent AID review (Ward et al., 1983, [TAB A1], p. 2) had this capability for doing mock-ups of long-term experience through short-term experiments. Nevertheless, such testing is invaluable in ascertaining maintenance requirements and systems life. In many cases economic viability is very close to being equal to the durability of the system, and then such tests will be crucial.

Evaluation has been defined as the end result of monitoring. Evaluation measures progress toward planned targets, examines unplanned results, determines causality, and separates internal elements from external factors that affect project performance, according to one review (Hammond, 1979, p. 7). Overall, the evaluation should be able to suggest any redesign necessary to improve the project. Various types of evaluation schemes can be valid: the most important point is to ensure the efficient use of scarce resources by measuring effectiveness in achieving planned objectives.[17]

17. Four different kinds of evaluations have been identified (Turner, 1979, pp. 76-77). One is a *formative evaluation*, when the project purpose is ill-defined and the managers need to formulate strategies to focus the objectives. The second is a *summative evalution* that measures progress toward a well-understood objective. The third is *goal attainment evalution*, when the project has a single dominant objective, and the fourth is a *systems evaluation*, when institution-building or projects with multiple objectives are considered.

Evaluations may be carried out at various bureaucratic levels. Routine management evaluations can be done at intermediate or regional levels. Evaluations made during the program may be decentralized at the project team level, while the ex post evaluations after project completion are normally carried out at a centralized program and policy level (Soedjatmoko, 1979, p. 61).

Specialized evaluations for particular purposes may also be useful. For example, for AID projects in Africa, surveys of user experience with systems, equipment, and tools under actual field conditions have been recommended as exceedingly useful in estimating maintenance costs and durability of systems (Ward et al., 1983, p. 63).

The most important principle is that of precision, pertinence, and practicality. The evaluation needs to be complete and should be timed to meet the specific needs of the key decision makers (Turner, 1979, p. 73). But most important, the system must be used: it has been noted that the "existence of a monitoring and evaluation system did not necessarily mean project personnel were using it" (Ward et al., 1983, p. B-42).

Conclusion and Planning Procedure: Bioenergy on a National Scale

The bioenergy planner has the special problem that his work may involve several different agencies. When these agencies have a narrow view of their role, the task of the bioenergy planner will be made more difficult, because it will be harder to gain necessary cooperation in such matters as land use, water allocation, and competition for scarce labor and management factors.

Opportunities arise for making use of the heterogeneity of the economic development process by combining bioenergy goals with other compatible objectives, however.

Going from a single project to a large program is a formidable step. The usual bioenergy planning problem at the local level does not take into account the full range of this problem. In fact, this problem more properly belongs at the national planning level. The local planner, however, cannot usually escape from the need to assess the "project versus program" problem as it affects him. The bioenergy planner must be aware, for example, how the specific project would interface with the potential demand of his country for biomass energy or for competitive conventional fuels to find out how the project would stand up as part of a larger program. The question of liaison with local governments and communities also becomes critically important under these circumstances; the interaction with local institution building can be crucial. Finally, a well-designed plan for monitoring and evaluation becomes essential if the project is to be considered as a small but significant piece of a larger programmatic picture.

15

Planning Procedures and Suggestions

Introduction: Plans and New Directions

Planning for bioenergy in bioenergy cultivation and conversion and its interactions with other development problems can be systematized into an overall planning procedure for project siting and assessment. The research carried out to develop this bioenergy guide has also stimulated some suggestions for further research and policy actions.

Review of the Planning Procedure

Planning bioenergy projects, like planning for any kind of economic development program, must take into account a great number of factors. The topics treated in each chapter of this book give a review of the factors that will connect the usual cost and engineering specifications of candidate bioenergy projects to the wider needs of bioenergy feedstock production and conversion—and to the still wider requirements of the national economic development process. To tie this process together, most of the chapters above contain a section headed "Planning Procedure." We now recap the steps in that procedure to show how a systematic approach to developing plausible bioenergy project options should be approached.

The goal of first considering national energy demand patterns is to answer the key question: does the bioenergy project being considered constitute an option that will show promise of making significant contributions to future energy needs (Chapter 2)? Both bioenergy conversion schemes and projects for bioenergy cultivation or other enhancement or conservation (e.g. use of wastes) of the bioenergy resource must answer this question. In this context, the nonrenewable use of biomass resources has obvious drawbacks; "mining" the forests is no long-term answer. Next, existing land uses will have to be examined to find and evaluate plausible siting areas—especially for biomass feedstock schemes, which tend to be land-intensive (chapter 3). In practice, the a priori choice of land will be restricted by political, legal, or other constraints. Land use data are generally not as available as one would like; in addition, it is difficult to specify site suitability in scientific terms. But indirect evidence, like present vegetation patterns, may give valuable guidance.

For biomass feedstock projects, the next step is a consideration of cropping choices, or the question of species selection or identification of usable wastes (chapter 4). In the present state of knowledge, field trials are still essential for many crops. In addition, because of the many unknown factors involved, the biomass project options involving cultivation must be investigated for possible long-term effects that may not show up in shorter trials; for forest crops the role of uncertainties such as seed quality is also important. Sociological factors affect the use of certain kinds of feedstocks (e.g., wood quality questions). For wastes, possible alternative uses must also be included in the plausibility assessment. The final assessment question involving the crop site itself is the environmental problem. Some help can be obtained here by examining soil and rainfall characteristics to make some predictions on erosion and by keeping in mind the special problems posed by loose granular soils, for example.

If the project involves conversion technologies, a main task will be to match up the conversion technologies to the supply of feedstock (chapter 6). If the conversion option is not tied directly to a separate cultivation or harvesting scheme, then some of the factors considered in the preceding chapters must be reviewed as they affect feedstock availability. Factors such as nonavailability of water in semi-arid regions for irrigated cultivation and the existence of unutilized weeds that could serve as bioenergy feedstocks would affect conversion decisions. The choice of conversion technology would also be closely tied into the questions of demand considered in chapter 2. For example, supplying liquid fuels is one of the biggest priorities in many countries, but there are numerous exceptions, especially in countries depending on thermal sources of electricity. The environmental dimension to bioenergy conversion facilities should to a large extent be a question of adapting project design to desirable environmental standards, including the use of effluents as byproducts when feasible (chapter 7). Environmental concerns will, however, directly influence some option assessments such as in giving environmental credit for the clean-burning qualities of biogas.

Bioenergy planners will have to perform their own assessments of the cost and engineering feasibility of specific bioenergy options based on data from proposals and project papers. In addition to the cost factors usually treated, however, they will want to consider the long-term outlook for costs and finances (chapter 8). Long-term land trends tend to be paramount, but projected patterns of competition from alternative fuels are also of concern. The role of the project in the foreign exchange balance, usually taken into account in project costing, is nevertheless a topic that is exceedingly important in examining tradeoffs for a number of countries with large external debts. The social complexities of the questions of scale also deserve investigation.

Infrastructure, distribution, and markets pose problems that may cause otherwise promising bioenergy options to be reconsidered (chapter 9). It is all too easy to assume the presence of support institutions that do not in fact exist—especially for questions like credit availability or transportation as they affect sometimes peripheral but important links for the project with the outside world.

A key part of the planning procedure is to check out whether the type of organizational format chosen is both appropriate and feasible—as considered in chapters 10, 11, and 12. Options to develop large-scale commercial facilities may come up against dubious economics in some cases and against lack of transferability of entrepreneurial skills in some byproduct operations. Community management can produce future social benefits, but it can also be hamstrung by existing social deficiencies. Whether an option uses these organizational approaches, or approaches the problem from the point of view of the individual or project-specific associations or other organizational formats, the examination of the option in the light of previous experience is all-important.

In one sense, the planning procedure ends at this point—all that is left is the place (usually modest) of the project in the national economy. In practice, however, options that open up lines of energy development that appear to impinge on other economic development goals may constitute undesirable candidates (chapter 13). Furthermore, the option must be proof both against neglect on the one hand and overenthusiasm on the other as far as government policy toward the use of biomass energy is concerned (chapter 14). Finally, an individual project can be and often is considered as strictly a one-shot phenomenon. Unless it can be consistent with the development of larger programs that produce important amounts of bioenergy, however, it may have little relevance for the developing country involved (chapter 14).

This systematization of the problem of assessing biomass energy conversion or feedstock supply options is designed to supplement the usual descriptions of engineering and costs that are available in proposals and in the literature for various bioenergy options. Ideally, given minimal uncertainties in data and a better knowledge of biological and economic laws, such an analysis could be modeled to produce optimal solutions to the bioenergy planning problem. In fact, given the poor data and deficient knowledge of causal relationships available, a qualitative consideration of the key factors in the system is the only practical approach. The planning procedure presented here suggests the outlines of such an approach.

Suggestions for Further Action and Research

This book is primarily a guide designed to help orient planning procedures for bioenergy projects in the context of a rational economic development strategy. Although the object is therefore not to make recommendations, some suggestions for further action and research appear to arise naturally from the results of the review of the data carried out in preparing the guide.

For example, in the arena of energy demand, the data problems common to so many other fields have been noted. Nevertheless, planners have to deal with insufficient data every day. And because most of the important decisions have to do with the intrinsically unknowable future, it is not clear whether it is either useful or relevant to devote resources to clarifying fine points in the demand data. At any rate, given the obvious impacts of the energy crisis on many countries, ignorance of exact energy requirements should certainly not excuse delaying attempts to find new solutions—like the bioenergy option.

The data problem in land use is somewhat different. Here, a more comprehensive data set is essential if expensive mistakes are going to be avoided while selecting crops for cultivation. The *cost* of acquiring better land use data may be the most significant barrier. The idea of cooperating on data gathering programs with other development programs therefore needs to be stressed, because national land use planning programs can usually better justify the expense of such tools as satellite imagery and side-looking aerial radar than can an isolated bioenergy plan.

For species selection decisions related to biomass cropping, it is evident that a great deal of scientific talent is going into the testing of new varieties. Could this work benefit from a more focused approach? It is certainly conceivable that some kind of *triage* could be useful in encouraging research teams to concentrate on a limited number of species and varieties and establish their ranges, flexibilities, and the values of other important parameters. Perhaps research in plant genetic engineering or breeding could also be focussed to produce a few varieties now, rather than hundreds in 10 years time. In the related environmental arena, it might well be that more concentrated work on a few common soil types could help establish the suitability of a wide number of sites for a number of existing and exotic crops.

Although the question of suitability of feedstocks is usually treated under the rubric of bioenergy conversion technologies, special research efforts on feedstock suitability might pay great dividends. In related environmental controls, the biggest money saver would probably be to include controls in plans for new facilities, rather than to have to retrofit after environmental concerns become more fully expressed—as they surely will be as development proceeds.

In the matter of scale of production, an interesting question is whether or not the "Small Is Beautiful" philosophy is still alive and well. In more operational terms, how much success has the intermediate technology movement had so far according to its own standards and/or according to conventional economic standards? Finally, the infrastructure problem is well understood by most planners, especially the question of roads. The subject of repairs and maintenance, however, is somewhat lacking in glamour. Has this crucial factor been treated as a stepchild is planning assessments?

When it comes to the organization of bioenergy projects, it appears that large commercial ventures—at least in the absence of a program of government intervention in the Brazilian or Chinese style—are usually not yet viable. A major exception might be in the area of by-products. A good deal of work has been done in this area, but a detailed review of industries in which bioenergy could be a by-product is needed. In the community bioenergy area, the greatest danger appears to be that of reinventing the wheel: what sort of effort has been made to see that different countries and agencies profit by each other's experience?

An examination of all of the other ways of organizing bioenergy projects seems to point out the importance of an active government role. One "sleeper" possibility must be the use of government-owned land or government-run institutions: more attention should be paid to this area. One of the newest organizational ideas that looks promising is that of associations, voluntary or government-sponsored, of smallholders to cultivate or process biomass for energy.

One conclusion that can be inferred from a good deal of the material reviewed here is that bioenergy may have an important role in integrated development schemes. Forestry programs, in fact, have traditionally involved multiple products. In the future, the use of biomass for energy may provide a stimulus not only to new agricultural schemes such as sugarcane production or fish culture but also to the development of industries such as plastics factories using as feedstocks residues from bioenergy production.

The situation with distribution and marketing is somewhat unsatisfactory. Part of the problem is that a great many assumptions have been made of how marketing and distribution in traditional bioenergy markets work, and hard information is difficult to come by. If future markets have to be organized or regulated by governments and not left to the marketplace itself, then experience in the centrally planned economies will be more relevant than is usually realized.

In theory, the interaction of bioenergy programs with national planning is an exceedingly important topic. In practice, although the conclusions here are based on fragmentary data, it seems likely that this issue is not as crucial as it is sometimes made out to be. Naturally there will be pressure on land prices and on production and prices and costs of other competing goods. But the assumption implicit in the phrase "food versus fuel" is often that biofuels are bad because they might compete with foodcrops. The hidden implication here—that such alternatives to biofuels as imported oil offer a free lunch to a developing society—is rather astounding in the context of any serious economic analysis. Erosion control on a country-wide basis is of course important, and bioenergy has to help rather than add to that problem; but for the average national planner this is a question of preaching to the converted. Bioenergy can affect equity and land tenure, but what, in fact, does not? It is not clear that alternative energy sources are any better—or any worse. Examining the question of how countries, national policies, and institutions may affect the development of bioenergy is interesting, if only because it points out how little one knows about the mechanics of national programs. It also reveals how little analytic—as opposed to descriptive—evidence there is on important large national programs such as those in Korea, Tanzania, and Brazil.

The question of "projects versus programs" could probably benefit from an analysis of additional data on existing projects Despite the brevity of the treatment given here, it is probably safe to hazard a guess that while small may be good, thinking small can be disastrous: the planner must keep the big picture in mind if his project is to make some real contribution to bioenergy in the context of economic development.

Bibliography

Adeyoju, 1976 — Adeyoju, D. Kolade. "Land Use and Tenure in the Tropics," *Unasylva*, vol. 28, nos. 112-113 (1976), pp.26-41.

AGA, 1983 — American Gas Association. "An Economic Efficiency and Environmental Comparison of Alternative Vehicular Fuels: 1983 Update," *Energy Analysis* (Arlington, VA: American Gas Association, September 9, 1983).

Ahn, 1978 — Ahn, Bong Won. "Village Forestry in Korea," discussion paper presented at the Eighth World Forestry Congress, Djakarta, Indonesia, 16 October 1978.

AID, 1982 — U.S. Agency for International Development. "Philippines: Rural Energy Development Project," Project Paper #492-0375, Annexes I and II (Washington, D.C.: U.S. Agency for International Development, 1982).

AID, 1983a — U.S. Agency for International Development. "Costa Rica: Wood Gasifier and Town Electrification," information sheet from conference (Washington, D.C.: U.S. Agency for International Development, April 1983).

Alam et al., 1982 — Alam, Manzoor S., Joy Dunkerley, K.N. Gopi, and William Ramsay with Elizabeth Davis. "Fuelwood Survey of Hyderabad," draft report (Washington, D.C.: Resources for the Future, 1982).

Alcohol Week, 1982 — *Alcohol Week*, vol. 3, no. 27 (5 July 1982).

Alexander, 1982 — Alexander, Alex. "Tropical Biomass: A Resource For All Seasons," paper presented at Symposium on Fuels and Feedstock, San Juan, Puerto Rico, 26-28 April 1982.

Allen, 1984	Allen, Julia C. "Soil Response to Forest Clearing in the United States and the Tropics: Geological and Biological Factors," *Biotropica* (in press, Summer 1984).
Allen and Barnes, 1981	Allen, Julia C. and Douglas F. Barnes. "Deforestation, Wood Energy and Development," draft report (Washington, D.C.: Resources for the Future, 1981).
Allen and Davis, 1983	Allen, Julia C. and Frank Davis. "Environmental Effects of Bioenergy Projects," informal paper (Washington, D.C.: Resources for the Future, 1983).
Araya, 1981	Araya, Zerai. "Village Forestry in Tanzania," report to the Forest Division of the Ministry of Natural Resources and Tourism (Dar es Salaam, Tanzania: January 1981).
Bahadur, 1984	Bahadur, Shahyad. Planning Department, Utter Pradesh, India. personal communication (1984).
Barnes et al., 1983	Barnes, Douglas F., Julia C. Allen and William Ramsay. "Social Forestry," Research Paper D-73D (Washington, D.C.: Resources for the Future, April 1982).
Beasley, 1972	Beasley, R.P. *Erosion And Sediment Pollution Control* (Ames, Iowa: Iowa State University Press, 1972).
Benneman, 1980	Benneman, John R. "Energy from Aqua- culture Biomass: Fresh and Brackish Water Aquatic Plants," in *Energy From Biological Processes*, vol. 3, part B, section IV (Washington, D.C.: Office of Technology Assessment, September 1980).
Bhatia, 1981	Bhatia, Ramesh. "Fuel Alcohol from Agro-Products in India: A Study of Crop Substitutions, Food Prices and Employment," (Delhi, India: Institute of Economic Growth, 1981).
Black, 1968	Black, C.A. *Soil-Plant Relation* (New York: John Wiley & Sons Publishing Co., 1968).

Bostid, 1981 — Board on Science and Technology for International Development. *Energy for Rural Development: Renewable Resources and Alternative Technologies for Developing Countries* (Washington, D.C.: National Academy Press, 1981).

Bostid, 1982 — Board on Science and Technology for International Development. *Diffusion of Biomass Energy Technologies in Developing Countries* (Washington, D.C.: National Academy Press, 1982).

Braunstein et al., 1981 — Braunstein, Helen M., Paul Kanciruk, R. Dickinson Roop, Frances E. Sharples, Jesse S. Tatum, and Kathleen M. Oakes with Frank C. Kornegay and Thomas E. Pearson. *Biomass Energy Systems and the Environment* (New York, N.Y.: Permagon Press, 1981).

Brewbaker, 1980 — Brewbaker, James L. "Giant Leucaena (Koa Haole) Energy Tree Farm: An Economic Feasibility Analysis for the Island of Molokai, Hawaii," (Honolulu, Hawaii: Hawaii Natural Energy Institute, September 1980).

Brokensha and Riley, 1978 — Brokensha, David and Bernard Riley. "Forest, Foraging, Fences and Fuel in a Marginal Area of Kenya," paper prepared for the U.S. Agency for International Development Africa Bureau Firewood Workshop, Washington, D.C., 12-14 June 1978.

Brookhaven, 1980 — Brookhaven National Laboratory. *Senegal Food and Energy Study: Energy Use and Opportunities for Energy-Related Improvements in the Food System*. (Upton, New York: Brookhaven National Laboratory, August 1980).

Bungay, 1982 — Bungay, Henry R. "Overview of New Biomass Industries," in *Priorities for International Development: Proceedings of a Workshop* (Washington, D.C.: National Academy Press, 1982.)

Burley and Wood, 1976	Burley, J. and P.J. Wood, eds. *A Manual on Species with Provenance Research with Particular Reference to the Tropics*, (London: University of Oxford Press for the Commonwealth Forestry Institute, 1976).
Campbell, 1980	Campbell, Gabriel, J. "Outstanding Social Issues in the Proposed Madhya Pradesh Social Forestry Project," (Washington, D.C.: U.S. Agency for International Development, 1980).
Campos-López and Anderson, 1983	Campos-López, Enrique and Robert J. Anderson. *Natural Resources and Development in Arid Regions* (Boulder, Colorado: Westview Press, 1983).
Carioca, 1981	Carioca, Jose Osvaldo B, "Ethanol: The Brazilian experience and Strategies," available from the Georgetown University Center for Strategic and International Studies (February, 1981).
Carpenter, 1981	Carpenter, Richard A. *Assessing Tropical Forest Lands: Their Suitability for Sustainable Uses* (Dublin, Ireland: Tycooly International Publishing Ltd., 1981).
Carr, 1978	Carr, Marilyn. *Appropriate Technology for African Women.* (New York: United Nations Press, 1978).
Castberg et al., 1981	Castberg, A. Dedrick, Tetsuo Miyabara and Louis J. Goodman. "Hawaii Bagasse Project: United States," Chapter 3, in *Goodman and Love*, 1981.
Cecelski and Ramsay, 1979	Cecelski, Elizabeth and William Ramsay. "Prospects for Fuel Alcohols from Biomass in Developing Countries," in UNITAR, 1981.
Cecelski et al., 1979	Cecelski, Elizabeth, Joy Dunkerley and1979 William Ramsay. *Household Energy and the Poor in the Third World.* (Washington, D.C.: Resources for the Future, 1979).

Celis et al., 1982 — Celis, Rafael, Rosario Domingo, Luis Herrera, Mario Vedova and Juan Villasuso. "The Foreign Trade Deficit and the Food Crisis: Anticipated Results of an Aggressive Program of Alcohol Fuel Production in Costa Rica," (Washington, D.C.: Resources for the Future, 1982).

Cernea, 1980 — Cernea, Michael M. "Land Tenure Systems and Social Implications of Forestry Development Programs," (Washington, D.C.: U.S. Department of Agriculture and Rural Development, October 1980).

Chan, 1981 — Chan, George. "Integrated Biogas Development: Fiji," chapter 5 in Goodman and Love, (1981).

Chen & Chen, 1979 — Chen, Guang-Qian and Ming Chen. "The Development of Chinese Biogas," in UNITAR, 1981.

Choong, 1981 — Choong, Lee Peng. "Forest Land Classification in Malaysia," in Carpenter, 1981.

Chopra, 1984 — Chopra, K. Department of Non-Conventional Energy Sources, New Delhi, India. personal communication (1984).

Coates et al., 1982 — Coates, J.F., Henry Hitchcock and Lisa Heinz. "Environmental Consequences of Wood and Other Biomass Sources of Energy," (Washington, D.C.: J.F. Coates, Inc., April 1982).

Collins, 1982 — Collins, G.B. "Plant Cells and Tissue Culture: An Overview," in *Priorities in Biotechnology Research for International Development: Proceedings of a Workshop* (Washington, D.C.: National Academy Press, 1982).

Conitz, 1982 — Conitz, Merril. "Regional Remote Sensing Facility," in *Proceedings of Workshop on Energy, Forestry and Environment.* (Washington, D.C.: U.S. Agency for International Development April 1, 1982).

Data Resources, Inc., 1983 — Data Resources Inc. *Review of the U.S. Economy.* (Washington, D.C.: Data Resources, Inc., March 1983).

De Beijer, 1979	De Beijer, I.R. "How Biomass Harvesting as Part of Logging Can Yield Highly Competitive Fuel Source," *World Wood* (June, 1978), pp. 27-29.
Denton, 1983	Denton, Frank H. *Wood for Energy and Rural Development; The Philippine Experience.* (Manila: Frank Denton, 1983).
Desai, 1981	Desai, Ashok V. "Interfuel Substitution in the Indian Economy," Discussion Paper D-73B (Washington, D.C: Resources for the Future, July 1981).
Dias, 1979	Dias, Celiodivo G. "Fuel-Alcohol Program: A New Brazilian Miracle," informal paper for School of Advanced International Studies (Washington, D.C.: Johns Hopkins University, Spring 1979).
Dobbs et al., 1982	Dobbs, Thomas L., Randy Hoffman and Ardelle Lundeen. "Evidence on Economic Feasibility of Small Scale Fuel and Alcohol Production," (Brookings, South Dakota: South Dakota State University, 1982).
DOE, 1979	U.S. Department of Energy. "Environmental Readiness Document: Biomass Energy Systems," (Washington, D.C.: U.S. Department of Energy, September 1979).
DOE, 1980	Department of Energy. "An Environmental Assessment of the Use of Alcohol Fuels in Highway Vehicles," (Washington, D.C: Center for Transportation Research, October 1980).
DOE, 1982	U.S. Department of Energy. "Assessment of Methane-Related Fuels for Automotive Fleet Vehicles," 3 Vols., Report No. DOC/CE 50179-1 (Washington, D.C.: U.S. Department of Energy, February 1982).
Dunkerley, 1980	Dunkerley, Joy. *Assessment of Energy Demand for Selected Developing Countries.* (Washington, D.C.: Resources for the Future, 1980).

Dunkerley et al., 1981 — Dunkerley, Joy, William Ramsay, Lincoln Gordon and Elizabeth Cecelski. *Energy Strategies for Developing Nations* (Baltimore: Johns Hopkins University Press for Resources for the Future, 1981).

Dunkerley, 1982 — Dunkerley, Joy. *Future Energy Consumption in India, Brazil, the Republic of Korea and Mexico.* (Washington, D.C.: Resources for the Future, 1982).

Dunkerley and Ramsay, 1983 — Dunkerley, Joy and William Ramsay. With Caroline Bouhdili and Elizabeth Davis, *Analysis of Energy Prospects and Problems of Developing Countries.* (Washington, D.C.: Resources for the Future, 1983).

Dunne and Leopold, 1978 — Dunne, T. and L.B. Leopold. *Water in Environmental Planning* (San Francisco, California: W.H. Freeman & Co., 1978).

Duvigneaud and Denaeyer-de Smet, 1970 — Duvigneaud, P., and S. Denaeyer-de Smet. "Biological Cycling of Minerals in Temperate Deciduous Forests," in *Analysis of Temperate Forest Ecosystems* Edited by D.E. Reichle (New York: Springer-Verlag, 1970).

Eckhaus, 1977 — Eckhaus, Richard S. *Appropriate Technologies for Developing Countries* (Washington, D.C.: National Academy of Sciences, 1977).

Elachi, 1982 — Elachi, Charles. "Radar Images of the Earth from Space," *Scientific American*, vol. 247, no. 6 (December 1982), pp. 54-61.

Elliot et al., 1978 — Elliot, L.F., J.M. McCalla, and A. Waiss. "Phytotoxicity Associated with Residue Management," *Crop Residue Management Systems* (Madison, Wisconsin: American Society of Agronomy, 1978).

Elwell, 1978 — Elwell, H.A. "Modeling Soil Losses in South Africa," *Journal of Agricultural Engineering Research*, vol. 23 (1978), pp. 117-127.

Evans, 1982 — Evans, Julian. *Plantation Forestry in the Tropics* (New York: Oxford University Press, 1982).

Fagundes Netto, 1980 — Fagundes Netto, Fernando. "Alternative Energy Sources: The Brazilian Approach," Speech given in Denver, Colorado, July 1980.

FAO, 1976a — Food and Agricultural Organization of the United Nations. "Harvesting Manmade Forestry in Developing Countries," (Rome: Food and Agriculture Organization, 1976).

FAO, 1981 — Food and Agriculture Organization of the United Nations. *1980 FAO Production Yearbook*, vol. 34 (Rome: Food and Agriculture, 1981).

FAO, 1981a — Food and Agriculture Organization of the United Nations. *Tropical Forest Resources: Asia* (Rome: Food and Agriculture Organization, 1981).

FAO, 1982 — Food and Agriculture Organization of the United Nations. *Village Forestry Development in the Republic of Korea* Report No. GGP/INT/347/SWE (Rome: Food and Agriculture Organization, 1982).

FAO, 1982a — Food and Agriculture Organization of the United Nations. *Agricultural Residues: Compendium of Technologies* (Rome: Food and Agriculture Organization, 1982).

FAO, 1983a — Food and Agriculture Organization of the United Nations. *Wood Fuel Surveys* Report No. GCP/INT/365/SWE (Rome, Food and Agriculture Organization, 1983).

Fege et al., 1979 — Fege, Anne S., Robert E. Inman and and David Salo. "Energy Farms for the Future," *Journal of Forestry* vol. 77, no.6 (June 1979), pp. 358-361.

Fenton et al., 1977 — Fenton, R.T., R.E. Roper and G.R. Watt. "Lowland Tropical Hardwoods—An Annotated Bibliography of Species with Plantation Potential," (New Zealand: Ministry of Foreign Affairs, 1977).

Flavell, 1981 — Flavell, Richard B. "An Overview of the Role of Plant Genetic Engineering in Crop Production," paper delivered at the Rockefeller Conference on Genetic Engineering For Crop Improvement, 12-15 May 1980 (New York: Rockefeller Foundation, August 1981).

Foley and Moss, 1983 — Foley, Gerald and Patricia Moss. *Improved Cooking Stoves in Developing Countries*. Technical Report No. 2 (London: Energy Information Programme, 1983).

FPL, 1980 — Forest Products Laboratory, U.S. Department of Agriculture Forest Service. "Forestry Activities and Deforestation Problems in Developing Countries," report to the Office of Science and Technology Development Support Bureau (Washington D.C: U.S. Agency for International Development, June 1980).

Goodman and Love, 1981 — Goodman, Louis J. and Ralph N. Love, eds. *Biomass Energy Projects: Planning and Management* (New York: Pergamon Press, 1981).

Goodrich, 1982 — Goodrich, Robert S. "PROALCOOL: Brazil's Program for Alcohol Motor Fuel," (Birmingham, Alabama: Southern Research Institute, 1982).

Gordon, 1979 — Gordon, Lincoln. *Growth Policies and International Order: 1980s Project/Council on Foreign Relations* (New York: McGraw Hill Book Company, 1979).

Greaves and Hughes, 1976 — Greaves, A. and J.F. Hughes. "Site Assessment in Species and Provenance Research," in Burley and Wood, 1976

Green, 1978 — Green, Maurice B. *Eating Oil: Energy Use in Food Production* (Boulder, Colorado: Westview Press, 1978).

GRI, 1982 — Gas Research Institute. "*Marine Biomass Program Annual Report: January-December 1982*," (Chicago, Ill.: Gas Research Institute, 1982).

GRI, 1983 — Gas Research Institute. "Baseline Project of U.S. Energy Supply and Demand for the Years 1982-2000," (Chicago, Ill.: Gas Research Institute, 1983).

Guellec, 1980 — Guellec, J. "Possibilités d'utilisation d'image landsat améliorées, a l'échele de 1/200.00, pour la connaissance des forêts," in *Bois et Forêts des Tropiques*, no. 193 (Septembre-Octobre 1980).

Guyer, 1984 — Guyer, David L. "Women in Development: Looking to the Future," Testimony submitted to the Senate Committee on Foreign Relations, Hearing on Women in Development. 98th Cong., 2d sess. (Washington, D.C.: no publisher, June 7, 1984).

Hall et al., 1982 — Hall, D.O., G.W. Barnard and P.A. Moss, *Biomass for Energy in Developing Countries* (Oxford: Pergamon Press, 1982).

Hall, 1983 — Hall, David. "Home-grown Fuel: Rise of Gasohol Challenge," *South* (December 1983) pp. 78-79.

Hamilton, 1981 — Hamilton, Lawrence S. "Some 'Unbiased' Thoughts on Forest Land Use Planning—from an Ecological Point of View," in Carpenter, 1981.

Hammer, 1982 — Hammer, Turi. "Reforestation and Community Development in the Sudan," CEPR Discussion Paper D-73M (Washington, D.C.: Resources for the Future, September 1982).

Hartshorn, 1979 — Hartshorn, Gary. Letter to Peter B. Martin. Report No. GSG-12 Jari (Hanover, N.H.: Institute for Current World Affairs, September 1979).

Hertzmark, 1979 — Hertzmark, Donald. "A Preliminary Report on the Agricultural Sector Impacts of Obtaining Ethanol from Grain," (Golden, Colorado: Solar Energy Research Institute, July 1979).

Hertzmark et al., 1980 — Hertzmark, Donald, Daryl Ray and Gregory Parvin. *The Agricultural Sector: Impacts of Making Ethanol from Grain* (Golden, Colorado: Solar Energy Research Institute, March 1980).

Heybroek, 1982 — Heybroek, Hans M. "The Right Tree in Right Place," *Unasylva* vol. 39, no. 136 (1982), pp. 15-19.

Hira et al., 1983 — Hira, Ayub U., Joseph A. Mulloney, Jr. and Gregory J. D'Alessio. "Alcohol Fuels from Biomass," *Environment, Science and Technology*, vol. 17, no. 5 (1983), pp. 202A-213A.

Hirschman, 1958 — Hirschman, Albert O. *The Strategy of Economic Development* (New Haven: Yale University Press, 1958).

Hirschman, 1967 — Hirschman, Albert O. *Development Project Observed* (Washington, D.C.: Brookings Institution, 1967).

Holdridge, 1981 — Holdridge, Leslie R., "World Life Zone Systems," in Carpenter, 1981.

Hoskins, 1979 — Hoskins, Marilyn W. "Community Participation in African Fuelwood Production, Transformation, and Utilization." discussion paper prepared for Overseas Development Council/U.S. Agency for International Development Workshop on Fuelwood and Other Renewable Fuels in Africa, Paris, November 1979.

Hoskins, 1980 — Hoskins, Marilyn W., *Social Dimensions in Land Forestry/Conversion Efforts*, (Washington, D.C.: World Bank, 1980).

Hoskins, 1982 — Hoskins, Marilyn W. "Social Forestry in West Africa: Myths and Realities," report presented at the annual meeting of the American Association for the Advancement of Science, Washington, D.C., 8 January 1982.

Howe, 1977 — Howe, James, W., "Energy for Villages of Africa: Recommendations for African Government Outside Donors," (Washington, D.C.: Overseas Development Council, February 25, 1977).

Hughes and Willan, 1976 — Hughes, J.F. and R.L. Willan. "Policy, Planning, and Objectives," in Burley and Wood, 1976.

Hyman, 1982 — Hyman, Eric L. "Pulpwood Treefarming for the Viewpoint of the Smallholder: An Ex Post Evaluation of the PICOP Project," draft, revised version (Philippines: no publisher, 1982).

Hyman, 1982a — Hyman, Eric L. "Analysis of the Woodfuels Market: A Survey of Fuelwood Sellers and Charcoal Makers in the Province of Ilocos Norte, Philippines," (Rome: Food and Agriculture Organization, 1982).

Hyman, 1983 — Hyman, Eric L. "Analysis of the Demand for Fuelwoods by Rural and Urban Households in the Province of Ilocos Norte, Philippines," draft (Philippines: no publisher, 1983).

IEA, 1983 — International Energy Agency. *Energy Balances of OECD Countries: 1971-1981*. (Paris: Organization for Economic Cooperation and Development, 1983).

Islam and Ramsay, 1982 — Islam, Shafiqul and William Ramsay. "Fuel Alcohol: Some Economic Complexities in Brazil and in the United States," Discussion Paper D-73G (Washington, D.C.: Resources for the Future, April 1982).

Islam et al., 1983 — Islam, Nural M., Richard Morse and M. Hadi Soesastro. *Rural Energy to Meet Development Needs: Asian Village Approaches*, (Honolulu, Hawaii: East-West Center, September 1983).

Java, 1983 — Java, R.L. "Development of Community Forestry Project," informal paper (Gujarat, India: Government of Gujarat, India, 1983).

Java, 1984 — Java, R.L. "Integrating Biomass Energy Programs," (Gujarat, India: Government of Gujarat, India, 1984).

Java, 1984a — Java, R.L. Conservator of Forest and Director of Training, Research and Communication, Government of Gujarat, India. personal communication (1984).

Jensen et al., 1977 — Jensen, Homer, L.C. Graham, Leonard Porcello and Emmett N. Leith, "Side-looking Airborne Radar," *Scientific American* vol. 237, no. 4 (October, 1977), pp. 84-95.

Jordan, 1977 — Jordan, Carl F. "Distribution of Elements in a Tropical Montane Rain Forest," *Tropical Ecology*. vol. 18, no. 2 (1978), pp. 124-130.

Jordan, 1978 — Jordan, Carl F. "Distribution of Elements in a Tropical Montane Rainforest," *Tropical Ecology*. vol. 18, no. 2 (1978), pp. 124- 130).

Kjellstrom, 1981 — Kjellstrom, Bjorn. *Producer Gas 1980: Local Electricity Generation from Wood or Agricultural Residues*. (Stockholm, Sweden: The Beijer Institute, 1981).

Knowland and Ulinski, 1979 — Knowland, Bill and Carol Ulinski, "Traditional Fuels: Present Data, Past Experience, and Possible Strategies," report for U.S. Agency for International Development (Washington, D.C.: U.S. Agency for International Development, September 1979).

Kockenderfer and Aubertin, 1975 — Kockenderfer, J.N. and G.M. Aubertin. *Municipal Watershed Management, Symposium Proceeding* (Washington, D.C.: U.S. Department of Agriculture, 1975).

Korea, 1981 — Republic of Korea Planning Bureau of the Ministry of Energy and Resources and the U.S. Department of Energy. "Republic of Korea/United States Cooperative Energy Assessment," 3 vols. (Argonne, Ill.: Argonne National Laboratory, September 1981).

Kozlowski, 1962 — Kozlowski, Theodore, ed. *Tree Growth* (New York: Ronald Press Co., 1962).

Kristoferson and Kjellstrom, 1981 — Kristoferson, Lars and Bjorn Kjellstroem. "Energy Provision for the Seychelles: Producer Gas and Other Opportunities," (Stockholm: Beijer Institute, September 1981).

KVIC, 1978 — Khadi and Village Industries Commission. "Gobar Gas: Retrospect and Prospects," (Bombay, India: Khadi and Village Industries Commission, 1978).

KVIC, 1983 — Khadi and Village Industries Commission, "Biogas: A Renewable Source of Energy: Prime Minister Visits Masudpur Biogas Complex." Supplement to *Biogas Newsletter*, no. 6, (Bombay, India: Khadi and Village Industries Commission, January 26, 1983).

KVIC, 1983a — Khadi and Village Industries Commission, "Gobar Gas: Why and How," (Bombay, India: Khadi and Village Industries Commission, March 1983).

KVIC, 1983b — Khadi and Village Industries Commission, "Community Biogas Project: Mehrauli, New Delhi," (Bombay, India: Khadi and Village, 1983).

Lee, 1957 — Lee, Douglas H.K. *Climate and Economic Development in the Tropics* (New York: Harper and Brothers, 1957).

Lele, 1974 — Lele, Uma. *The Design of Rural Development: Lessons from Africa* (Baltimore: John Hopkins University Press for the World Bank, 1974).

Likens and Borman, 1978 — Likens, G.E. and F.H. Bormann. "Linkages Between Terrestrial and Aquatic Ecosystems," *Bioscience*, vol. 24 (1974), pp. 447-456.

Lockeretz, 1982 — Lockeretz, William, ed. *Agriculture as a Producer and Consumer of Energy* (Boulder, Colorado: Westview Press for the American Association for the Advancement of Science, 1982).

Lovins, 1977 — Lovins, Armory B. "Scale, Centralization and Electrification in Energy Systems," (Oak Ridge, TN: Oak Ridge Associated Universities, 1977).

Mahin, 1982 — Mahin, Dean B. "Bioenergy Systems: Biogas in Developing Countries," (Washington, D.C: U.S. Agency for International Development, March 1982).

McPherson, 1984 — McPherson, M. Peter. Testimony submitted to the Senate Committee on Foreign Relations, Hearing on Women in Development. 98th Cong., 2d sess. (Washington, D.C.: no publisher, June 7, 1984).

Meta, 1980 — Meta Systems Inc. "Potential For Fuelwood and Charcoal in the Energy Systems of Developing Countries," input for panel at United Nations Conference on New And Renewable Sources of Energy (Cambridge, Mass.: Meta Systems Inc., 1980).

Meta, 1982 — Meta Systems, Inc. "An Examination of the Substitution of Woody Based Fuels for Oil in the Industrial Sectors of Costa Rica," report prepared for the Instituto Tecnologico de Costa Rica/Citizen's Energy Corporation Program (Cambridge, Mass: Meta Systems, Inc., May 1982).

Meyers and Jennings, 1979 — Meyers, Henri and R.F. Jennings. "Charcoal Ironmaking: A Technical and Economic Review of the Brazilian Experience," (New York: United Nations Industrial Development Organization, 2 November 1979).

MITRE, 1981 — Mitre Corporation. *Bioenergy Handbook for Developing Countries* draft report for the Agency for International Development, Bureau for Science and Technology, Office of Energy (Washington, D.C.: Agency for International Development, December 2, 1981).

Morgan and Icerman, 1981 — Morgan, Robert P. and Larry J. Icerman. *Renewable Resource Utilization for Development*. (New York: Pergamon Press, 1981).

Moss, 1975 — Moss, Michael R. "Biophysical Land Classification Schemes: A Review of Their Relevance and Applicability to Agricultural Development in the Humid Tropics," *Journal of Environmental Management* vol. 3, no. 4 (October 1975), pp. 287-307.

Moulik and Srivastava, 1975 — Moulik, T.K. and U.K. Srivastava. *Bio-gas Plants At the Village Level: Problems and Prospects in Gujarat* (Ahmedabad, India: Indian Institute of Management, 1975).

Mueller-Dombois, 1981 — Mueller-Dombois, Dieter. "The Ecological Series Approach to Forest land Classification," in Carpenter, 1981.

NAS, 1975 — National Academy of Sciences. "Underexploited Tropical Plants with Promising Economic Value," (Washington, D.C.: National Academy of Sciences, 1975).

NAS, 1977 — National Academy of Sciences. *Leucena: Promising Forage and Tree Crop for the Tropics* (Washington, D.C.: National Academy Press, 1977).

NAS, 1980 — National Academy of Sciences. "Firewood Crops: Shrub and Tree Species for Energy Production," (Washington, D.C: National Academy of Sciences, 1980).

NAS, 1983 — National Academy of Sciences "Alcohol Fuels: Options for Developing Countries," (Washington D.C: National Academy of Sciences, 1983).

Ndour, 1983 — Ndour, Mme. Country Paper for Senegal presented at the A.I.D. Energy Analysis, Planning and Policy Development Conference, Reston, Virginia, 27 February- 4 March 1983.

Nelson, 1973 — Nelson, Michael. *The Development of Tropical Lands: Policy Issues in Latin America* (Baltimore: John Hopkins University Press for Resources for the Future, 1973).

Netschert, 1983 — Netschert, Bruce. National Research Associates, slide presentation, Washington, D.C., 19 May 1983.

Netto, 1984a — Netto, Odwaldo Bueno, Jr. Consultant to Georgetown Center for Strategic and International Studies, Washington, D.C. personal communication (1984).

Netto, 1983 — Netto, Odwaldo Bueno, Jr. Consultant to Georgetown University Center for Strategic and International Studies, Washington, D.C. personal communication (November, 1983).

Nikles et al, 1978 — Nikles, D.G. J. Burley and R.D. Baines. *Progress and Problems of Genetic Improvement of Tropical Forest Trees*, report for the Commonwealth Forestry Institute (London: University of Oxford Press, 1978).

Noronha, 1980 — Noronha, Raymond. "Village Woodlots: Are They A Solution?" (Washington, D.C.: National Academy of Sciences, 1980).

NRC, 1981 — National Research Council. *Food, Fuel, and Fertilizer from Organic Wastes* (Washington, D.C.: National Academy Press, 1981).

NRC, 1982 — National Research Council. "Plant Cell and Tissue Culture," in *Priorities in Biotechnology Research for International Development: Proceedings of a Workshop* (Washington, D.C.: National Academy Press, 1982).

NRC, 1983 — National Research Council. *Calliandra: A Versatile Small Tree for the Humid Tropics* (Washington, D.C.: National Academy Press, 1983).

Nye, 1981 — Nye, P.H. "Organic Matter and Nutrient Cycles under Moist Tropical Forest," *Plant and Soil*, vol. 13, no. 4 (1961), pp. 333-346.

OLADE, 1981 — Organizacion Latinoamericana para Desarrollo y Energia. *Energy Balances for Latin America* (Ecuador: OLADE Printing Office, 1981).

OTA, 1980 — Office of Technology Assessment. *Energy from Biological Processes*, 3 vols. (Washington, D.C.: U.S. Government Printing Office, 1980).

Pelin, 1980 — Pelin, Eli. "Elevacão do Preço da Terra e Deslocamento de Culturas," (São Paulo, Brazil: University of São Paulo, 1980).

Philips, 1970 — Philips Research Laboratories. "How the Two Photosystems Came to Light," *Scientific American* vol. 223, no. 3 (September, 1970), p. 147.

Piarulli, 1984 — Piarulli, Susan. "The Role of Women in Bioenergy: Ignore at Your Own Risk," draft (Washington, D.C.: Center for Strategic and International Studies, 1984).

Pimentel et al, 1982 — Pimentel, David, Caren Fried, Lars Olson, Steven Schmidt, Kathy Wagner-Johnson, Anne Westman, Anne Marie Whelan, Kim Foglia, Peter Poole, Ted Klein, Rodney Sobin and Amy Bochner. "Environmental and Social Costs of Biomass Energy," draft (1982).

Poole, 1983 — Poole, Alan. "The Impact of the Alcohol Program on the Agricultural System of Brazil," report for the U.S Agency for International Development (Washington, D.C.: Resources for the Future, 1983).

Powell, 1978 — Powell, John W. *Wood Waste as an Energy Source in Ghana* (Boulder, Colorado: Westview Press, 1978).

Ramsay and Anderson, 1972 — Ramsay, William and Claude Anderson. *Managing The Environment: An Economic Primer* (New York: Basic Books, Inc., 1972).

Ramsay, 1978 — Ramsay, William. "Cost Prospects for Renewables in Developing Countries," presentations at International Seminar of Energy, Santo Domingo, Costa Rica, 4-7 January 1978.

Ramsay, 1979 — Ramsay, William. *Unpaid Costs of Electrical Energy* (Washington, D.C.: Resources for the Future, 1979).

Ramsay, 1979a — Ramsay, William. "Prospects for Biomass Fuels in Developed Areas: Methanol and Other Thermo-chemical Conversion Products," (Washington, D.C: Resources for the Future, July 10, 1979).

Ramsay, 1980b — Ramsay, William. "Alcohol Fuels and Developing Countries," unpublished manuscript, 1980.

Ramsay and Shue, 1981 — Ramsay, William and Elizabeth Shue. "Infrastructure Problems for Rural New and Renewable Energy Systems" *Journal of Energy and Development*, vol. 6, no. 2 (March, 1981), pp. 232-250.

Ramsay and Jankowski, 1982 — Ramsay, William and John E. Jankowski. "Alcohol Fuels and the Agricultural Sector." Discussion Paper D-73J (Washington, D.C.: Resources for the Future, 1982).

Ranganathan, 1981 — Ranganathan, Shankar. "Gujarat Forestry Scheme: The Start of Something Good," *Parks Magazine, 1981*, vol. 5, no. 4 (January-March, 1981), pp. 10-11.

Rask and Adams, 1979 — Rask, Norman and Reinaldo I. Adams. "Regional Competiveness of Alcohol Production in Brazil," paper presented at the 1979 National Meeting of the Latin Studies Association (Columbus, Ohio: Department of Agricultural Economic and Rural Sociology, Ohio State University, 1979).

Ratasuk, 1981 — Ratasuk, Sermpol, "Potential of Anaerobic Digestion in Biogas Production in Developing Countries." in UNITAR, 1981.

Reynolds, 1975	Reynolds, D.R. "An Appraisal of Rural Women in Tanzania," report for International Development Office of Women in Development (Washington, D.C.: U.S. Agency for International Development, December 1975).
Ribeiro and Branco, 1979	Ribiero, C. Costa and J.R. Castello Branco. "Ethanol Stillage: A Resource Disquised as a Nuisance," paper presented to the Third Bi-Annual International Conference on Effluent Treatment in the Biochemical Industries, Rio de Janeiro, Brazil, 6-7 November 1979.
ROCAP, 1979	ROCAP. "Fuelwood and Alternative Energy Sources," Project Papter AID/LAC/P: 031, for the U.S. Agency for International Development (Washington, D.C.: U.S. Agency for International Development, 1979).
Rockwood, 1983	Rockwood, Walt. "New Biotechnology In International Agricultural Development," *Horizons*, vol. 2, no. 10 (November, 1984), pp. 21-27.
Roose, 1977	Roose, E.J. "Use of the Universal Soil Loss Equation to predict Erosion in West Africa," in *Soil Erosion: Prediction and Control* (Ankeny, Iowa: Soil Conservation Society of America, 1977).
Russell, 1973	Russell, E. Walter. *Soil Conditions and Plant Growth* (London: Longman Press, 1973).
Samantha and Sundaram, 1983	Samantha, B.B. and A.K. Sundaram. "Socio-economic Impact of Rural Electrification on India," Discussion Paper No. D-730, (Washington, D.C.: Resources for the Future, 1983).
Schipper et al., 1982	Schipper, Lee, Jack Hollander, Matthew Milukas, Joseph Alcomo and Stephen Meyers. "Energy Conservation in Kenya's Modern Sector: Progress, Potential and Problems," Discussion Paper D-731 (Washington, D.C.: Resource for the Future, May 1982).

Schmithuesen, 1976	Schmithuesen, Franz. "Forest Utilization Contracts on Public Lands in the Tropics," *Unasylva*, vol. 28, nos. 112-113 (Rome: Food and Agriculture Organization, 1976) pp. 52-73.
Schumacher, 1973	Schumacher, E.F. *Small is Beautiful: Economics as if People Mattered* (New York: Harper & Row, 1973).
Schurr et al., 1979	Schurr, Samuel H., Joel Darmstadter, Harry Perry, William Ramsay, Milton Russell. *Energy in America's Future: The Choices Before Us* (Baltimore: Johns Hopkins University Press for Resources for the Future, February 1980).
Science, 1983	*Science*, vol. 219, no. 4569, (October 1983).
Sedjo, 1979	Sedjo, Roger. "Forest Plantations in Brazil," draft (Washington, D.C.: Resources for the Future, November 1979).
SERI, 1981	Solar Energy Research Institute. (Golden, Colorado: Solar Energy Research Institute, April 1981).
Shaikh and Larson	Shaikh, Asif and Patricia Larson. "The Economics of Village-Level Forestry: A Methodological Framework," report to the U.S. Agency for International Development Africa Bureau (Washington, D.C.: Agency for International Development, no date).
Silva, 1982	Silva, Paulo Robert. "Agricultura e Producão de Energia no Nordeste do Brasil," Working Paper No. 13 for the University of Arizona College of Agriculture (Tuscon, Arizona: University of Arizona, 1982).
Singh, 1977	Singh, I. "Tanzania: Some Empirical and Policy Considerations." in *Basic Economic Report, Annex VII: Appropriate Technologies in Tanzanian Agriculture* (New York: World BanK, December 1977).
Skutsch, 1983	Skutsch, Margaret McCall. "Why People Don't Plant Trees: The Socioeconomic Impacts of Existing Woodfuels Program: Village Case Studies, Tanzania." CEPR Discussion Paper D-73P (Washington, D.C.: Resources for the Future, March 1983).

Skutsch, 1984 — Skutsch, Margaret McCall. Lecturer, Development Studies, Technology and Development Group, Twente University of Technology. personal communication, 14 August 1984.

Smale et al., 1984 — Smale, Melinda, Michelle Savoie, Zahra Shirwa Cabdi and Moxamed Cali Axmed. "Cooking Practices and Fuelwood Consumption in Selected Sites of Lower Shabelle, Banaadir and Gedo Regions," (Arlington, Virginia: Volunteers in Technical Assistance, February 1984).

Smil, 1979 — Smil, Vaclav. "Renewable Energies: How Much and How Renewable," *Bulletin of Atomic Scientists* vol. 35 (December 1979), pp. 12-19.

Smith and Aggarwal, 1983 — Smith, Kirk R. and Dave R.M. Aggarwal. "Air Pollution and Rural Biomass Fuels in Developing Countries: A Pilot Village Study in India and Implications for Research and Policy," in *Atmospheric Improvement,* vol. 17, no. 11 (1983).

Smith, 1984 — Smith, Elise Fiber. "Women and Development: Looking Toward the Future," Testimony submitted to the Senate Committee on Foreign Relations, Hearing on Women in Development (Washington, D.C.: no publisher, June 7, 1984).

Soedjatmoko, 1979 — Soedjatmoko. "National Policy Implications of the Basic Needs Model, "in *Development Digest.* vol. XVII, no. 3 (July 1979), pp. 55-68.

Spears, 1978 — Spears, John S. "Wood as an Energy Source," Speech given at the 103rd Annual Meeting of the American Forestry Association, Hot Springs, Arkansas, 8 October, 1978.

Spears, 1980 — Spears, John S. "Developing Country Forestry Issues in the 1980s." (Washington, D.C: World Bank, November 1980).

Streets, 1962 — Streets, R.J. *Exotic Forest Trees in the British Commonwealth* (Oxford: Clarendon Press, 1962).

Sudan, 1982 — National Education Association and U.S. Agency for International Development. "Bioenergy for the Sudan." report to the Bureau of Science and Technology (Washington, D.C.: U.S. Agency for International Development, July 1982).

Taylor, 1981 — Taylor, Robert P. "Rural Energy Development in China," (Washington, D.C: Resources for the Future, 1981).

Terrado, 1981 — Terrado, Ernesto H. "Biogas Development: The Philippines," chapter 5, in Goodman and Love, 1981.

Thrupp, 1981 — Thrupp, Lori Ann, "Food, Fuel, and Forestry: Renewable Resources Programs in Latin America." (Washington, D.C: Resources for the Future, 1981).

Thrupp, 1983 — Thrupp, Lori Ann. "Women, Wood, and Work: The Imperative for Equity in Overcoming a Deeper 'Energy Crisis'," informal paper (Surrey, England: Institute of Development Studies, April 1983).

Trindade, 1984 — Trindade, Sérgio C. President of Intercor, Rio de Janeiro, Brazil. personal communication, 1984.

Turner, 1979 — Turner, Herbert D. "Program Evaluation in AID: Some Lessons Learned," *Development Digest*. Vol. XVII, No. 3 (July, 1979), pp. 71-79.

Umali, 1981 — Umali, Ricardo M. "Forest Land Assessment and Management for Sustainable Uses in the Philippines," in Carpenter, 1981.

UN, 1980a — United Nations. "Policies and Programmes of the United Nations Fund for Population Activities in the Field of Women, Population and Development," report prepared for the World Conference of the United Nations Decade for Women, Copenhagen, Denmark, 14-30 July 1980.

UN, 1983 — United Nations. *Energy Balances 1977-1980 and Electricity Profiles 1976-1981 for Selected Developing Countries and Areas.* (New York: United Nations Printing Office, October 1983).

UNITAR, 1981 — United Nations Institute for Training and Research. *Long-Term Energy Resources* (Marshfield, Mass.: Pitman Publishing Inc., 1981).

U. of Costa Rica, 1981 — Universidad de Costa Rica, "El Alcohol en Costa Rica," (Costa Rica: Universidad de Costa Rica, September 1981).

Van Buren, 1979 — Van Buren, Ariane. "The Chinese Development of Biogas and its Applicability to East Africa," paper presented at the Energy and Environment in East Africa Conference, Nairobi, Kenya, May 1979.

Velez, 1983 — Velez, Eduardo. "Rural Electrification in Columbia," (Washington, D.C.: Resources for the Future, March 1983).

Waldbott, 1978 — Waldbott, G.L. *Health Effects of Environmental Pollutants* (St. Louis: C.V. Mosby Company, 1978).

Ward et al., 1983 — Ward, Mark, John H. Ashworth, and George Burrill. "Renewable Energy Technologies in Africa: An Assessment of field Experience and Future Directions," report for the U.S. Agency for International Development (Washington, D.C.: U.S. Agency for International Development, December 1983).

Watt, 1982 — Watt, G.R. "A Renewable Energy Source: Wood," *Energy Exploration and Exploitation,* vol. 1, no. 3 (1982).

Weatherly, 1983 — Weatherly, Paul. "The 'Do-ability' of Bioenergy," paper presented at Bioenergy Approaches in National Development Conference, sponsored by U.S. Agency for International Development, Reston, Va., March 1983.

Webb et al., 1980	Webb, Derek B., Peter J. Wood, and Julie Smith. *A Guide to Species Selection for Tropical and Sub-Tropical Plantations*, report for Commonwealth Forestry Administration, (London: University of Oxford, 1980).
Wilkinson, 1984	Wilkinson, Jack. Chief Economist, Sun Oil Co., Philadelphia, Pa., personal communication (1984).
Winterbottom, 1980	Winterbottom, Robert T. "Reforestation in the Sahel: Problems and Strategies." African Studies Association Annual Meeting, Philadelphia, Pa., 15-18 October 1980.
Wionczek et al., 1982	Wionczek, Miguel S. and Gerald Foley, and Ariane Van Buren, eds. *Energy in the Transition from Rural Subsistence* (Boulder, Colorado: Westview Press, 1982).
World Bank, 1980	World Bank. *Alcohol Production from Biomass in Developing Countries* (Washington, D.C.: World Bank, September 1980).
World Bank, 1980a	World Bank. Personal communication: staff (1980).
World Bank, 1980b	World Bank. "Forestry Project Case Studies: Gujarat Community Forestry Project," draft (Washington, D.C.: World Bank, 1980).
World Bank, 1980d	World Bank. "Sociological Aspects of Forestry Project Design," AGR Technical Note No. 3 (Washington, D.C.: World Bank, November 1980).
World Bank, 1981	World Bank *Global Energy Prospects* Working Paper No. 489 (Washington, D.C.: World Bank, 1981).
World Bank, 1982d	World Bank, personal communication: staff (1982).
World Bank, 1982e	World Bank. "Guidelines for the Presentation of Energy Data in Basic Reports," (Washington, D.C.: World Bank, October 1982).
World Bank, 1982g	World Bank, personal communication: staff (1982).
World Bank, 1982i	World Bank. "Bangladesh: Rural and Renewable Energy Issues and Prospect," Energy Department Paper, No. 5 (Washington, D.C.: World Bank, 1982).

World Bank, 1982h — World Bank. personal communication: staff (1982).

World Bank, 1982j — World Bank. *World Bank Development Report: 1982*. (Washington, D.C.: World Bank, 1982).

World Bank, 1982y — World Bank, personal communication: staff (1982).

World Bank, 1982z — World Bank, personal communication: staff (1982).

World Bank, 1983 — World Bank. *World Bank Development Report: 1983*, (Washington, D.C: World Bank, 1983).

World Bank, 1983a — World Bank. "The Energy Transition in Developing Countries," Report #4442 (Washington, D.C.: World Bank, 1983).

Wourou and Tran, 1982 — Wourou and Van Nao Tran. "Orienting Forestry Towards the Needs of the People," *Unasylva*, vol. 34, no. 136, (1982), pp. 8-10.

Wunsch, 1980 — Wunsch, James. "Renewable Resource Management, Decentralization and Localization in the Sahel: The Case of Afforestation," (Omaha, Nebraska: no publisher, 9 October 1980).

Index

Acacia saligna, 79, 80–81(tables), 86–87(table)
Acacia senegal, 80–81(tables), 86–87(table), 193, 205, 210
Acid hydrolysis, 113, 125
Adobe, 66
Aerial photography, 33, 48, 49
Afforestation, 198. *See also* Reafforestation
Afghanistan, 40–46(table)
Africa, 240, 244, 246
  energy consumption, 23(table), 24, 26, 27(table)
  land use, 40–46(table), 188, 191
  species, 79
  See also Labor, women; *individual countries*
Agency for International Development (AID), 49, 119, 144, 207, 240, 244, 245, 246. *See also* Women in Development
Agricultural export industries, 26, 219
Agriculture. *See* Agroforestry; Biomass, competition with food crops; Land use
Agroforestry, 100–101, 155, 185, 204, 206, 219, 253
AID. *See* Agency for International Development
Albizia procera, 85, 86–87(table)
Alcogas program (Philippines), 169
Alcohol, 1, 2, 9, 15, 51, 53, 54, 60, 66, 67, 104, 106–107, 108, 112–114, 116, 117, 144, 162–163, 217
  hydrated, 146
  See also Ethanol; Methanol; PROALCOOL
Aldehydes, 128
Alder. *See Alnus* species
Alexander, Alex, 85
Alfalfa, 59
Algae, 63, 98, 160(n15)
Algeria, 40–46(table)
Alnus species, 82
Amino acid, 83
Andhra Pradesh. *See* Hyderabad
Anerobic digestion, 125–126. *See also* Biogas, production
Anglican Church (Tanzania), 207
Angola, 40–46(table)
Annual Crop Production Plan (Brazil), 171
Antigua, 40–46(table)
Appliances, 24, 25–26, 110, 126, 208–209, 242
Appropriate technologies, 148, 154–155
Appropriate Technologies International (ATI), 209
Aquatic crops, 31(n1), 62–63, 69, 98–99, 116
Arable land, 32, 40–46(table). *See also* Land use
Argentina, 23, 24, 27(table), 40–46(table), 159(n14), 168
Arid-land cultivation, 5, 79, 80–81(tables), 116, 250
Arizona, 62
Ash, 120, 121, 124, 132
Ash (tree). *See Fraxinus*
Asia, 23(table), 27(table), 40–46(table), 64, 66. *See also individual countries*
Associations, 168, 169, 198, 199, 202–203, 210, 211, 240
ASTRA. *See* Center for the Application of Science and Technology in Rural Areas
ATI. *See* Appropriate Technologies International
Australia, 64, 79, 168
Avon Products, Inc., 192

Bacteria. *See under* Soil
Bagasse, 22, 26, 54, 61, 66, 91, 109, 112, 116, 124, 160, 162, 170, 174–175
Baghouses, 121
Bahamas, 40–46(table)
Bahrain, 40–46(table)
Balance of payments. *See* Foreign exchange
Bamboo, 59
Bananas, 67
Bangladesh, 8, 25, 40–47(tables), 199

Barbados, 40–46(table)
Bark, 58, 160
Belize, 40–46(table)
Benin, 40–46(table), 156(n3), 183(n8)
Bermuda, 40–46(table)
Bermuda grass, 101
Bhutan, 40–47(tables)
Biochemical oxygen demand (BOD), 124
Bioenergy
  blitz programs, 234–235
  conversion technologies, 1–3, 5, 15, 36, 68, 108–117, 119–133, 250, 253
  costs, 2–3, 59, 60, 64–65, 70, 103, 107, 116, 135–144, 149, 215–219, 227, 231, 238, 239, 250–251, 253. *See also* Economies of scale
  defined, 1, 51
  and environment, 91–101, 102, 119–133, 177, 221–223, 227, 238, 250
  industrial organization, 3, 56
  and infrastructure, 151–155, 160, 163, 236, 241, 251
  macroeconomic factors, 144–145, 149, 238–239
  marketing and distribution, 156–163, 169, 171, 240, 241, 251, 253–254
  and national policy, 229–231, 233–235, 238–247, 251
  planning, 1, 2, 3, 5, 10–11, 16, 31, 33–34, 35, 36–37, 48–50, 135–136, 151–152, 210–211, 213–227, 229–247, 249–254
  project replication, 239–242, 246
  projects, 1–2, 3, 31, 53, 59, 62, 66, 144–145, 152, 155, 160, 167–173, 197–198, 231–236, 239–246, 249–250
  projects, commercial and state, 165–176, 179, 198–202, 204, 210, 211, 232, 235–236, 251, 253, 254
  projects, community, 177–195, 199, 202–211, 233–234, 235–236, 251
  sociopolitical considerations, 145, 147–149, 177, 181, 213, 223–226, 227, 233, 235, 240–241, 243–244, 247, 250, 251
  See also Biogas; Biomass; Evaluation; Land use; Maintenance and repair; Monitoring
  Biofuels. *See* Biogas; Charcoal; Ethanol; Fuelwood; Methanol; Wastes
Biogas
  costs, 147, 161(n19)
  effluent, 160(n15), 182
  liquified, 130
  and pollution, 129–130, 176, 250
  production, 1, 66, 105, 108, 114–115, 116, 125–126, 142, 160, 176, 207–208, 231, 241–242
  projects, 161, 199, 207–208, 231, 235
  sites, 185
  uses, 15, 106, 117, 126, 162
  See also Gobar gas
Biogeochemical cycle, 92
Biomass, 15, 241
  byproducts, 62, 142, 160, 165, 170, 250, 253
  competition with food crops, 54, 61, 65, 185, 213, 215–216, 217, 220, 225, 226–227, 230, 232, 254
  consumption, 22, 23–30
  demand, 159, 250, 252
  energy per unit, 119
  farm, 167
  harvesting, 96, 98
  and income, 26. *See also* Equity
  and labor, 64. *See also* Labor
  per unit time, 57
  planting, 95, 187
  prices, 65, 90, 112, 114(n33), 171, 219, 235, 238
  production, 63–65, 67
  sites, 36–37, 64, 69, 92–96, 102, 184–185, 252
  sources, 1, 2, 5, 15, 26, 51–63
  species selection, 67–70, 71–82, 89, 104, 185, 250, 252
  uses, 1, 15, 23–30, 83, 84, 104–108, 160
  yield of burnable dry matter, 67–68, 82
Biomes, 33
Bio-oil, 107, 115
Blue-green algae, 63
BOD. *See* Biochemical oxygen demand
Bolivia, 23(table), 27(table), 40–46(table)
Botswana, 40–46(table), 140
Brazil, 9, 23(table), 27, 40–46(table), 54, 55(n8), 107, 112, 136, 143, 145, 146, 161, 162, 167–168, 174, 232, 253, 254
  land use, 217, 218, 219, 220, 225, 237
  See also PROALCOOL
Bricks, 66
Briquetting, 109
British Indian Ocean Territory, 40–46(table)

British Virgin Islands, 40–46(table)
Bromine, 62
Brunei, 40–47(tables)
Buffer strips, 101, 102
Building materials, 66, 69, 83, 160, 170, 203, 216, 237
Burma, 40–47(tables)
Burundi, 7, 40–46(table)

California, 62
Calliandra calothyrsus, 56(n9), 74–75(table), 85, 204
Cambodia. *See* Kampuchea
Cameroon, 40–46(table)
Cape Verde, 40–46(table)
Capital, 137–138, 143–144, 149, 239, 241. *See also* Credit
Carbohydrates, 231. *See also* Starch; Sugar
Carbon char, 132
Carbon dioxide, 62, 92, 114, 125, 223
Carbon monoxide (CO), 108, 112, 127, 129, 133
Carcinogens, 122, 127, 132, 133
Caribbean, 40–46(table)
Cash crops, 219
Cash markets, 11
Cassava, 61, 124, 146, 171, 220
Castor oil, 61, 105
Catalina Island (Calif.), 62
Cation exchange capacity, 96
Cattails. *See Typha*
Cayman Islands, 40–46(table)
Cellulose, 54, 58–60, 69, 96, 113, 125
CEMAT. *See* Centro Mesomerico de Estudios Sobre Tecnologia Apropriado
Center for the Application of Science and Technology in Rural Areas (ASTRA) (India), 147
Central African Republic, 40–46(table)
Central America, 24, 26, 40–46(table), 83. *See also individual countries*
Centrally planned economies, 180, 182
Centro Mesomerico de Estudios Sobre Tecnologia Apropriado (CEMAT), 209
Chad, 40–46(table)
Charcoal, 7, 15, 22, 25, 26, 27, 67, 83, 90, 105(n6), 106, 107, 108, 109(ns13, 15), 110–111, 116, 117, 121–122, 142, 157, 158, 159, 178
  pollution, 127
  production, 165, 166–167, 168, 174, 226
Char oils, 121
  slurries, 107
Chase Manhattan Bank, 192
Chicory, 61(n20)
Chile, 23(table), 27(table), 40–46(table)
China, 40–46(table), 62, 155, 176, 178, 180, 181, 182, 183(n9), 184, 208, 231, 235, 253
Chromosomes. *See* Genome
Circular loan system, 154
Climate, 34, 35, 38, 250
  modification, 223
  See also Biomass, species selection; Temperate climates; Tropics
Clonal plant propagation, 89
CO. *See* Carbon monoxide
Coal, 17, 22, 104, 109(n13), 110, 115, 121, 238
  consumption, 23(table), 24, 26, 27(table)
  prices, 139–140, 143
Coconut
  husks, 25
  oil, 61, 105, 116
Coffee, 7, 26, 67
Coffee tree residue, 26
Colombia, 23(table), 25(table), 26, 27(table), 40–46(table), 139
Columbia (space shuttle), 50
Community Forest Project (India), 183
Comoros, 40–46(table)
Conflict resolution, 232, 233
Congo, 40–46(table)
Conservation. *See under* Energy; Forests
Cooperatives, 154, 158, 168, 169, 175, 202, 203
COPLAMAR project (Mexico), 200
Coppice, 55, 58, 69, 79, 82, 101
Corn, 54, 58, 98, 124, 146, 169, 219
Costa Rica, 8, 40–46(table), 90, 108(n11), 136, 137, 143, 144, 146, 162, 170, 173, 204, 232, 237, 238
Cottage industries, 205
Cottonwood. *See Populus* species
Credit, 153–154
Cremations, 157(n8)
Crud formation, 107
Cuba, 40–46(table)
Cyprus, 40–46(table)

Dairy farm digesters, 207
Debt. *See* Foreign exchange
Decentralization, 243
Deforestation, 27, 91, 96, 97(n17), 156(n3), 157(n4), 223
Dendrothermal program (Philippines), 168–169, 202, 240
Desert reclamation, 199

Devaluation, 145
Diesel, Rudolph, 61
Diesel fuel, 8, 12, 14, 105, 116, 117, 237, 238
  alcohol blends, 128
  prices, 143
Digesters. *See* Anerobic digestion; Biogas, production; Dairy farm digesters
Direct combustion, 109–110, 116, 120–121, 130, 238
Distillation, 54, 60, 109, 124, 125
Division of labor, 186, 192
Djibouti, 40–46(table)
Dominica, 40–46(table)
Dominican Republic, 23(table), 27(table), 40–46(table), 49
Double cropping, 59
Drought-resistant species. *See* Biomass, species selection
Dung, 216. *See also* Wastes, animal

East Timor, 40–46(table)
Economies of scale, 145–147, 148, 149, 165, 183–184, 253
Ecuador, 40–46(table), 142, 180
Effluents, 99–100, 119, 120, 121, 123, 124, 125, 130, 160(n15), 222–223, 250
  waste water, 126
Egypt, 40–46(table), 136, 237
Electricity, 8, 12, 14, 20, 21, 22, 104, 117, 141, 160, 174, 175
  consumption, 23(table), 24, 25, 26, 27(table), 159(n13)
  prices, 143, 162
  thermal, 17, 108, 116, 250. *See also* Dendrothermal program
  See also Hydroelectricity
Electrostatic precipitators, 121
Elm. *See Ulmus pumida*
El Salvador, 40–46(table)
Emissions, 121–122, 123, 124, 127, 128, 129, 130, 132–133, 222
Energy
  balance, 5–6, 14, 17–22, 28–29(table)
  capital, 144
  commercial, 22, 26. *See also* Industrial sector
  conservation, 9–10
  consumption, 22, 23(table)
  demand, 5, 6–11, 13, 15, 30, 67, 252
  prices, 8, 9
  primary, 17
  renewable, 141. *See also* Bioenergy; *specific types*
  secondary, 17
  traditional, 22
Energy transformation sector, 17, 20, 21, 22, 104. *See also* Bioenergy, conversion technologies
Entrepreneurs. *See* Smallholders
Enzymes, 113
Equatorial Guinea, 40–46(table)
Equity, 224, 227, 254
Erosion, 92, 98, 254. *See also* Soil, erosion
Esters, 105
Ethanol, 104, 105(n3), 106, 112, 114(n33), 123–125
  pollution, 128–129, 130
  prices, 172
  storage, 161(n20)
"Ethanol Stillage: A Resource Disguised As A Nuisance" (Ribeiro and Branco), 130
Ethiopia, 40–46(table), 167
Eucalyptus, 2, 55(n8), 56(n10), 57, 64, 68(ns31, 33, 35), 72–77(tables), 80–81(tables), 82, 85, 86–87(table), 161(n21), 168, 201, 232
Euphorbia, 62, 201
Evaluation, 245–246, 247
Evapotranspiration, 95
Exotic species, 55–58, 68, 69, 98, 252
Export
  controls, 235
  earnings, 7, 215, 218
Extension services, 155, 194, 204
Exxon (company), 192

Factor tradeoffs, 64–65, 216, 219
Falkland Islands, 40–46(table)
FAO. *See* Food and Agriculture Organization
Farm Systems Development Corporation (FSDC) (Philippines), 152, 203
Fermentation, 51, 54, 66, 106, 112–113, 116, 123, 124, 125, 130
  anerobic, 105, 114, 124
Ferric oxide, 129
Ferric sulfide, 129
Fertilizer, 59, 61, 64, 66, 69, 70, 95, 98, 114, 124, 126, 130, 160(ns15, 16), 237
Fiji, 114(n37), 160
Fischer-Tropsch process, 115
Fish culture, 253
Flood control, 99
Flood-resistant species, 85
Food and Agriculture Organization (FAO), 32, 39, 48, 82, 172, 216
Food prices, 217, 218, 219–221

Foreign exchange, 9, 13, 14, 144, 145, 149, 214–215, 232–233
Forest Research Institute (FRI) (Korea), 209
Forest residue, 97–98
Forests, 14, 15, 32, 34, 39, 40–47(tables), 48, 52–53, 55–58, 63, 91, 97, 198, 234, 200–207, 250
  conservation, 142, 143
  See also Plantation forestry; Social forestry program
Formaldehyde, 129
France, 17–21
Fraxinus, 85
Freight-hauling, 8, 143
French Guiana, 40–46(table)
FRI. *See* Forest Research Institute
FSDC. *See* Farm Systems Development Corporation
Fuelwood, 1, 2, 7, 8, 11, 12, 14, 15, 22, 27, 55, 69, 79, 82, 83, 110, 116, 117, 120, 141, 142, 178
  collection, 156–158, 186, 225
  consumption, 24, 26, 106, 107
  pollution, 126–127, 132–133, 144
  prices, 162
  production, 33, 147, 165–166, 167, 174, 180–181, 199–202, 203, 204–206
  See also Dendrothermal program
Furfural, 160(n16)
Fusel oil, 160(n16)

Gabon, 40–46(table)
Gaillardia, 11–13
"Galool" (Somali cooperative), 158
Gambia, 40–46(table)
Gas. *See* Biogas; Liquid petroleum gas; Natural gas; Producer gas; Water gas
Gasification, 107–108, 111(n23), 121, 122–123, 130, 146, 152, 155, 202, 238
Gasifiers, 67, 105, 108, 111, 117, 160(n18), 173, 203
Gasohol, 54, 169
Gasoline, 8, 9, 14, 117, 237
  alcohol additive, 54, 104–105, 128, 169
  consumption, 12
  prices, 143
GDP. *See* Gross domestic product
Genetic engineering, 88–89, 252
Genome, 88
Genotype, 88
Geothermal energy, 141
Germany, 115
Ghana, 5, 40–46(table), 55(n8), 65
Gmelina species, 57, 71, 74–77(tables), 78, 86–87(table)
GNP. *See* Gross national product
Gobar gas, 142
Government programs, 179–180, 197, 198–202, 210, 229–231, 253. *See also* Bioenergy, projects, commercial and state
Grain alcohol. *See* Ethanol
Grand total energy requirements (GTER), 28–29(table)
Grasses, 58, 59, 100, 101, 185
Grazing. *See* Pastureland
Green revolution, 230
Grenada, 40–46(table)
Gross domestic product (GDP), 12
Gross national product (GNP), 7, 136, 138
Ground cover, 100
Ground surveys, 33, 48, 50
GTER. *See* Grand total energy requirements
Guadeloupe, 40–46(table)
Guam, 63, 83
Guatemala, 40–46(table), 208–209
Guayule, 33, 62
Guinea, 40–46(table)
Guinea-Bissau, 40–46(table)
Gujarat project. *See* India, forests
Gum arabic, 205, 206
Gur, 220
Guyana, 40–46(table)

Haiti, 14, 40–46(table), 143, 222
Haploids, 89
Hardwoods, 58, 68(n32), 82
Hawaii, 64, 66, 162, 175, 203
HC. *See* Hydrocarbons
Herbicides, 64, 97, 99, 100
Heybroek, Hans M., 85
"Hidden resources," 177, 210
Hilo Coast Processing Company (Hawaii), 175
Honduras, 40–46(table)
Hong Kong, 40–46(table)
Households
  biogas production, 207–208
  biomass use, 24, 25, 27–30, 110, 159, 208–209
  energy efficiency, 10
  fuel use, 8, 14, 24, 105–107, 132–133, 156, 159
  See also Labor, women
H_2S. *See* Hydrogen sulfide
Hughes Aircraft Company, 49

Husks, 25, 67
Hybridization, 89
Hyderabad (India), 165–167, 176
Hydrocarbons (HC), 62, 104, 123, 125, 127, 128, 129, 132, 133
  synthetic, 107
Hydrochloric acid, 125
Hydroelectricity, 8, 11, 14, 17, 20, 99, 141, 222
Hydrogen, 108, 112, 114, 125
Hydrogen sulfide (H_2S), 126, 129
Hydrological cycle, 92, 95
Hydrolysis. *See* Acid hydrolysis; Wood hydrolysis

Income distribution. *See* Equity
India, 8, 11, 14, 23(table), 24, 25, 26, 27(table), 142, 147, 154, 157, 181, 237
  biogas, 106, 142, 178, 179, 181, 182, 187, 199, 207, 233
  forests, 47(table), 55(n8), 71, 165–167, 180, 182, 183–184, 185, 201–202, 203, 207, 234
  land use, 40–46(table), 185, 188, 217–218, 220, 225
Individual farmers. *See* Smallholders
Indonesia, 23(table), 27(table), 142–143
  biogas, 207
  forests, 47(table), 56(n9), 64, 204
  land use, 40–46(table)
Industrial sector
  biomass use, 26–30, 104, 107–108, 159(n14), 167–168, 174, 175
  fossil fuel use, 159
Inflation, 144, 145, 149
Infrared, 49
Insecticides, 98
Insolation, 141
Institute for Sugar and Alcohol (Brazil), 171
Institutions, 206–207, 210, 211, 242–244, 247
Instituto de Recursos Naturales Renovables (RENARE) (Panama), 200
Integrated Service Associations (ISA) (Philippines), 203
Interest rates, 137–138
Interfuel substitution, 8, 215, 237
Intermediate technology, 240, 241
International donor projects, 198–199, 207, 231, 234, 242
Interplanting, 59, 68(n35), 193, 219
Inversion, 133
Iodine, 62
Ipil-ipil. See Leucaena species
Iran, 40–46(table)
Iraq, 40–46(table)
Iron smelting, 168
Iron sponge, 129
Irrigation, 116, 152, 160(n15), 186, 201, 220, 222, 232
  pump fuel use, 14, 15, 25, 155, 200, 202, 203
ISA. *See* Integrated Service Associations
Islam, 191
Israel, 40–46(table)
Ivory Coast, 23(table), 27(table), 40–46(table)

Jamaica, 40–46(table), 59
Japan, 62, 144, 160(n17)
Jari operation (Brazil), 174
Jerusalem artichoke, 61(n20)
Jet fuel, 12
Jojoba, 62
Jordan, 40–46(table)
Jute sticks, 25

Kampuchea, 40–47(tables)
Kelp, 51, 62
Kenaf, 59
Kenya, 10, 23(table), 27(table), 40–46(table), 55(n8), 141, 154, 159, 169, 193, 194, 204, 209
Kerosene, 8, 12, 14, 24, 25, 26, 106, 107, 238, 239
  prices, 142–143, 162
Khadi and Village Industries Commission (KVIC) (India), 154, 187, 207, 233
Khartoum (Sudan), 158
Kilns, 111, 155
Korea, 23(table), 24, 27(table), 40–46(table), 56(n9), 82, 110, 136–137, 143, 160(n17), 181, 182, 183(n9), 184, 201, 207, 209, 222, 232, 234, 235, 243, 254
Kudzu bark, 160
Kuwait, 40–46(table)
KVIC. *See* Khadi and Village Industries Commission

Labor
  availability, 225–226
  costs, 136–137, 149
  intensive, 148, 149, 226
  skilled, 185–187, 208, 240
  training, 155, 181, 208
  underutilized, 177, 211
  women, 181, 186, 189–195, 206, 225–226

Land clearing, 94–95
Land costs, 136, 226, 254
Landfills, 66
LANDSAT, 33, 48–50
Land tenure, 36, 181, 185, 187–188, 190–191, 225, 254
Land use, 1, 2, 15, 31–50, 104, 185, 199, 215, 216–218, 219, 230, 237, 249–250, 252
  frontier, 217, 221
  marginal, 217, 220
Lao, 40–47(tables)
Latin America, 23(table), 27(table), 40–46(table), 159(n13), 178(n1), 237. *See also individual countries*
Latin American Organization for Development and Energy (OLADE), 21, 22
Leaf-cutter ants, 78
Leaves, 25, 54, 58, 65
Lebanon, 40–46(table)
Leguminous bushes, 82
Lesotho, 40–46(table)
Lespedeza species, 56(n9), 82
Leucaena species, 55(n8), 56(n9), 57, 63–64, 72–73(table), 83–85, 86–87(table), 168, 169
Liberia, 40–46(table)
Libya, 40–46(table)
Lipids, 63
Liquidambar styraciflua, 85
Liquid petroleum gas (LPG), 12, 26(n4)
  synthetic, 107
Locust tree. *See Robinia pseudoacacia*
Lorena stove, 208–209
Lovins, Amory B., 147
LPG. *See* Liquid petroleum gas

Macau, 40–46(table)
Madagascar, 40–46(table)
Maintenance and repair, 154, 163, 240, 246, 253
Malawi, 23(table), 27(table), 40–46(table), 180, 243
Malaysia, 40–47(tables), 53, 69
Maldives, 40–46(table)
Mali, 40–46(table), 153(ns1, 2), 203
Malting. *See* Sprouting
Mangroves, 199
Manioc. *See* Cassava
Mannitol, 62
Marsh gas, 114
Martinique, 40–46(table)
Mauritania, 40–46(table)
Mauritius, 40–46(table)
Maya Farms (Philippines), 176
Mesquite, 51, 79, 80–81(tables), 199
Methane, 1, 2, 62, 66, 105, 108, 114, 115, 116, 117, 124, 125, 126
  pollution, 129–130
Methanol, 104, 112, 115, 121, 122, 123
  pollution, 128–129, 222–223
  toxicity, 128, 130
Mexico, 24, 33, 40–46(table), 200
Microwave radar, 50
Mimosine, 83
Mindanao (Philippines), 53
Mobil Oil (company), 192
Modern sector, 14
Molasses, 66–67, 175, 220
Mongolia, 40–46(table)
Monitoring, 244–245, 246, 247
  accelerated use tools, 245
Monoculture, 219
Montserrat, 40–46(table)
Morgan Guaranty Trust Company, 192
Morocco, 40–46(table), 200
Mozambique, 40–46(table)
Mushrooms, 160(n17)

Namibia, 40–46(table)
NAS. *See* National Academy of Sciences
NASA. *See* National Aeronautics and Space Administration
National Academy of Sciences (NAS), 82
National Aeronautics and Space Administration (NASA), 49
National Alcohol Commission (Brazil), 171
National Council of Women (NCW) (Kenya), 194
National Energy Agency (NEA) (Philippines), 168, 169, 202
National Petroleum Council (Brazil), 162, 171
"National Program of Alcohol Fuel Production" (Costa Rica), 144
Natural gas, 21, 22, 26, 106, 112, 114, 218(n2)
  artificial, 122
  consumption, 23(table), 27(table)
  reserves, 140
NCW. *See* National Council of Women
NEA. *See* National Energy Agency
Neem tree, 55(n8)
Nepal, 14, 40–47(tables), 156(n3), 20[illegible], 234, 243
Netherlands Antilles, 40–46(table)
New Community Movement (Korea), 232, 234
Nicaragua, 40–46(table)

Niger, 40–46(table), 181, 185
Nigeria, 40–46(table), 53, 140
Nitrates, 62
Nitric oxide, 132
Nitrogen, 55, 58, 59, 63, 66, 69, 84, 94, 96, 98, 114, 121, 126, 231
Nitrogen dioxide, 132
Nitrogen oxides (NO_x), 121, 125, 127, 129, 133
Nomadic herding, 32
NO_x. *See* Nitrogen oxides
Nuclear power, 17, 140
Nutrients. *See* Nitrogen; Phosphorus; Soil

OAS. *See* Organization of American States
OECD/IEA. *See* Organization for Economic Cooperation and Development/International Energy Agency
Offcuts, 65
Oil, 7, 8, 9, 11, 12, 13, 14, 15, 17, 22, 238, 254
  consumption, 23(table), 27(table), 28–29(table)
  prices, 65, 138–139, 142–143, 149
  products, 20, 21, 24, 142
  See also Shale oil
Oil palms, 61, 105
Okra, 59
OLADE. *See* Latin American Organization for Development and Energy
Oman, 40–46(table)
Ondol stove, 209
OPEC. *See* Organization of Petroleum Exporting Countries
Organic carbon, 96, 114
Organization for Economic Cooperation and Development/ International Energy Agency (OECD/IEA), 21
Organization of American States (OAS), 49
Organization of Petroleum Exporting Countries (OPEC), 144
Oxygen, 114, 125, 130
Ozone, 132

PAH (Polycyclic Aromatic Hydrocarbons). *See* Polycyclic organic matters
Pakistan, 25, 40–47(tables), 55(n8), 204, 207, 225
Panama, 40–46(table), 200, 236
Paper Industries Corporation of the Philippines (PICOP), 162, 206
Papua New Guinea, 7, 47(table), 68(n31), 176
Paraguay, 40–46(table)
Parastatal firms. *See* Bioenergy, projects, commercial and state
Particulates, 120, 121, 122, 124, 125, 127, 128, 132, 223
Pastureland, 39, 40–46(table), 185, 218, 225
Pea family, 55
Peanut
  oil, 105
  production, 119
  shells, 67
Pelletizing, 109, 175
Percolating, 126
Peru, 40–46(table), 186, 199, 200
Pesticides, 95, 99–100
Petrobras (Brazilian state oil company), 171
Pfizer (company), 192
Phenotype, 88
Philippines, 11, 23(table), 24, 27(table), 68(n31), 111(n20), 152, 155, 157, 162, 168–169, 176, 202–203
  biogas, 207, 208
  forests, 47(table), 53, 55(n8), 204, 206
  land use, 40–46(table)
Phosphorus, 62, 63, 64, 96, 98, 231
Photosynthesis, 51
Photovoltaic cells, 141
Phragmites, 63
PICOP. *See* Paper Industries Corporation of the Philippines
Pig iron, 159, 167, 168
Pine. *See Pinus*
Pinus, 51, 55(n8), 56(n10), 57, 64, 68(n32), 72–77(tables), 81(table), 86–87(table), 160(n17)
Planktonic microalgae, 63
Plantation forestry, 2, 53, 55–58, 63, 71, 97, 141, 147, 152, 161, 167–168, 172, 179
Plantation systems, 32, 47(table)
Plastics, 160(n16), 170, 253
Platanus species, 82
Poland, 139
Poles. *See* Building materials
Pollution. *See* Effluents; Particulates; Stillage; Water pollution
Polycyclic Aromatic Hydrocarbons (PAH). *See* Polycyclic organic matters

Polycyclic organic matters (POMs), 121, 125, 127, 132–133
POMs. *See* Polycyclic organic matters
Poplar. *See Populus* species
Population. *See* Rural population; Urban population
Populus species, 76–77(table), 82, 85
Potash, 124
Potassium, 97
Private Agencies for International Development, 192
PROALCOOL (Brazilian alcohol program), 162, 169, 170, 171–172, 215, 218, 235
Process heat. *See* Households; Industrial sector
Producer gas, 15, 105, 106, 110, 111, 112, 117, 122, 127, 178, 241
Prosopis. *See* Mesquite
Puerto Rico, 40–46(table), 59, 85
Pulping operations, 174
Pulverizing, 109(n13), 110
Pyrolysis, 110–111, 121–123
  pollution, 127
Pyrolytic oil, 90, 107, 110, 111, 117, 241

Qatar, 40–46(table)

Radar. *See* Microwave radar; Side-looking airborne radar
Railways, 153
Ranching, 32. *See also* Pastureland
Ratoon, 217
Reafforestation, 167, 168, 183, 207, 231–232, 235
Recycling, 125
Reeds. *See Phragmites*
Reforestation, 200, 201
Regional Remote Sensing Facility (Nairobi), 49
Regrowth. *See* Coppice
Remote sensing. *See* LANDSAT
RENARE. *See* Instituto de Recursos Naturales Renovables
Reservoirs, 99, 222
Residues. *See* Forest residues; Slurry; Toxic residues; Tree residues; Wastes
Reunion, 40–46(table)
Rice, 220, 232
  husks, 25
Roads, 152, 153, 155, 160–161
Robinia pseudoacacia, 82, 85
Romania, 63(n25)
Rubber trees, 62, 69
Runoff, 94, 96, 98, 101, 102, 223
Rural Electric Cooperatives (Philippines), 168, 169, 202
Rural population, 11, 12, 24, 25–26, 106, 137, 148, 149, 152, 154, 159(n13), 160–161, 177, 226. *See also* Bioenergy, projects, community
Rwanda, 40–46(table), 204, 237, 243

Sahel, 57(n14), 156(n3)
St. Helena, 40–46(table)
St. Kitts-Nevis, 40–46(table)
St. Lucia, 40–46(table)
St. Vincent, 40–46(table)
Saline-resistant species, 85
Salix species, 82, 85
Sanitation, 231
Sao Tome and Principe, 40–46(table)
Satellites, 33, 38, 48, 252
Saudi Arabia, 40–46(table)
Sawdust, 1, 65, 90
Sawmills, 65, 90, 147, 165, 174
Schumacher, E. F., 147
Scrubbers, 121, 123, 127, 129
Sediment, 99, 101, 221
Seedlings. *See* Biomass, planting/species selection; Plantation forestry
Self-Help Scheme (India), 182
Senegal, 23(table), 27(table), 40–46(table), 179, 181(n6), 186, 187, 194, 195
Sesame oil, 61, 105
Sesbania, 64, 71, 72–75(tables)
Sewage treatment, 125–126
Seychelles, 40–46(table)
Shale oil, 140
Shavings, 65
Shifting cultivation, 32
Short-rotation energy cropping, 64
Shuttle Imaging Radar (SIR-A), 50
Side-looking airborne radar (SLAR), 33, 50, 252
Sierra Leone, 40–46(table)
Siltation, 222
Silviculture, 53, 167. *See also* Forests
Singapore, 40–46(table)
SIR-A. *See* Shuttle Imaging Radar
SLAR. *See* Side-looking airborne radar
Sludge, 126
Slurry, 107, 182, 231
Smallholders, 197, 203–207
Smog, 129, 132
Social forestry program, 183–184, 186
Softwoods, 58
Soil, 34, 38, 85, 92–98, 101, 102, 126, 216, 252
  bacteria, 55, 84, 94, 101

erosion, 66, 94, 95–96, 100, 102, 198, 199, 221, 222, 250
Solar energy, 51, 54, 141
Solar radiation. *See* Insolation
Solomon Islands, 53
Somalia, 40–46(table), 158
Sorghums, 59, 60
South Africa, 55(n8), 115, 139
SOVEDA (Senegal), 194
SO_x. *See* Sulfur oxides
Soybeans, 61, 105
Special Drawing Rights, 145
Sprouting, 113
Sri Lanka, 40–47(tables)
Starch, 51, 52, 53–54, 60, 113, 124
Steam engines, 15, 105(n8)
Steel, 167
Stem grass, 101
Stillage, 124, 125, 130, 160(n16), 221
Stirling (steam) engine, 105(n8)
Stoves. *See* Appliances
Straw, 1
Subsidies, 142, 143, 162–163, 174, 177, 181, 235, 238, 239
Subsistence farming, 32
Sudan, 8, 40–46(table), 67(n29), 99(n22), 111(n21), 155, 158, 161, 181, 190, 191, 193, 194, 195, 199–200, 205, 210
Sugar, 51, 52, 53–54, 60, 113, 114, 124, 125
Sugar beets, 54
Sugarcane, 1, 2, 51, 53, 54, 58, 59, 60, 61, 69, 91, 98, 116, 124, 146, 170–171, 217, 219, 232, 253
prices, 171, 220
resprouted. *See* Ratoon
See also Alcohol; Bagasse; Gur
Sulfur, 62, 120, 121
Sulfuric acid, 113, 129
Sulfurous acid, 129
Sulfur oxides (SO_x), 128, 129, 133
Sunflowers, 105, 116
Sunn hemp, 59
Supervised Village Scheme (India), 182
Suriname, 40–46(table)
Swampland, 63
Swaziland, 36, 40–46(table), 140, 161(n21)
Sweden, 110, 168
Sweetgum. *See Liquidambar styraciflua*
Sweet potatoes, 61
Syndicate Bank (India), 154
Synthesis gas. *See* Water gas
Syria, 40–46(table)

Taiwan, 207
Tanzania, 40–46(table), 83, 155, 156(n3), 181, 184, 186, 187, 188, 189, 190, 207, 235, 254
Tars, 122, 123
Taxes, 168
Teak, 55(n8)
Technology. *See* Appropriate technologies; Bioenergy, conversion technologies; Intermediate technology
Tectona species, 57
TEFR. *See* Total energy form requirements
Temperate climates, 59, 60, 82
TER. *See* Total energy requirements
Terms of trade, 144
TFA. *See* Tree Farmers' Associations
TFC. *See* Total final consumption
"Thai bucket," 209
Thailand, 23(table), 27(table), 40–47(tables), 49, 159(n14), 207, 208
Threshold limit value (TLV), 123(n10), 129
Tidal energy, 141
TLV. *See* Threshold limit value
Tobacco, 26
Togo, 7, 40–46(table)
Topography, 34, 38
Total energy form requirements (TEFR), 28–29(table)
Total energy requirements (TER), 17–21
Total final consumption (TFC), 17, 20, 21
Toxic residues, 98, 99–100
Traditional sector, 14–15, 159, 165. *See also* Households
Transportation, 153, 190, 217
costs, 149, 158, 185
fuel, 12, 14, 104, 157, 158, 161, 184–185, 226
See also Bioenergy, and infrastructure; Freight-hauling
Trans-sterification process, 105(n4)
Tree Farmers' Associations (TFA) (Philippines), 168, 202, 240
Tree residues, 26, 66, 90
Triangle Sugar Estate (Zimbabwe), 172
Trinidad and Tobago, 40–46(table)
Triple cropping, 60
Tropics, 39, 60, 71–78, 82
Tunisia, 40–46(table)
Turkey, 40–46(table)
Turks and Caicos Islands, 40–46(table)
Twigs, 25, 65
Typha, 63

Uganda, 40–46(table)
Ulmus pumida, 85
UNDP. *See* United Nations Development Program
United Arab Emirates, 40–46(table)
United Nations, 21, 22. *See also* Food and Agriculture Organization
United Nations Development Program (UNDP), 172
United States, 112, 139, 144, 221
Universal Soil Loss Equation (USLE), 102
Upper Volta, 40–46(table), 172, 179(n3), 181, 191, 234
Uranium, 140
Urban population, 24, 26, 148, 157, 159(n13), 226
Uruguay, 40–46(table)
USLE. *See* Universal Soil Loss Equation
Usutu forests (Swaziland), 36

Vegetable oil crops, 61–62, 69, 105, 116
Venezuela, 23, 24, 40–46(table)
Vietnam, 40–47(tables)
Village Forestry Associations (Korea), 181, 201, 202, 234
Village Polytechnics (Kenya), 193
Villages. *See* Bioenergy, projects, community; Rural population
Volcanic eruptions, 223

Wastes, 7, 8, 11, 15, 26, 65–67, 70, 106, 117, 156, 174, 219
  animal, 2, 22, 25, 66, 106, 142, 176, 182, 216, 231
  crop, 2, 22, 25, 26, 66–67, 174, 216, 231, 237
  production, 33
  See also Waste water; Wood wastes
Waste water, 63, 123(n9), 126
Water gas, 108, 112, 115, 122, 127
Water hyacinths, 2, 63, 98, 116
Water pollution, 123, 124
Water transport, 153
Webb, Derek B., 79, 82
Weeding, 64, 95
Weeds, 57, 63, 78, 250
Western Sahara, 40–46(table)
Wheat, 54, 59, 98, 232
Willow. *See Salix*
Wind energy, 141, 238
Women. *See under* Labor
Women for Progress (Kenya), 209
Women in Development (AID), 192–193
Wood. *See* Fuelwood; Woodlots; Wood wastes
Wood alcohol. *See* Methanol
Wood hydrolysis, 125
Woodlots, 1, 24, 63, 160, 180, 181, 183–188, 190, 199, 203, 204, 235
Wood wastes, 65, 67, 90, 109, 162, 165
World Bank, 21, 22, 144, 199, 234
World Energy Conference (1974), 141

Yearbook of World Energy Statistics (UN), 21
Yeast enzymes, 113
Yemen, North and South, 40–46(table)

Zaire, 40–46(table)
Zambia, 40–46(table)
Zimbabwe, 7, 23(table), 24, 27(table), 40–46(table), 140, 169, 172